Clinical Decision Making

Croydon Health Sciences Library
Mayday University Hospita...
London Road
Thornton Heath
Surrey CR7 7YE

Clinical Decision Making: an Art or a Science?

E. J. Kay,* MPH, PhD, FDS RCPS
N. M. Nuttall,** BSc, PhD

Senior Lecturer in Health Services Research, Turner Dental School, University Dental Hospital of Manchester
***Senior Research Fellow, Dental Health Services Research Unit, Dental School, University of Dundee*

Published by the British Dental Association
64 Wimpole Street, London W1M 8AL

ISBN 0 904588 49 1

Printed and bound in Great Britain by Latimer Trend & Company Ltd, Plymouth

Preface

Every general dental practitioner knows that decision making is the lynch pin upon which rests the success of his or her practice. Practitioners also know that their relationships with patients and their self esteem depend on their ability to rapidly process information and then choose an appropriate action. However, despite the multitude of decisions made by dentists each day and the profound effect that these decisions have on the oral health of the nation, attempts to understand how such decisions are made are new. This is because judgement, intuition and experience all form part of the process of decision making which dental clinicians use each day. The focus of this book is on the outcome of the decisions we make, rather than on dissecting out component parts of the process. The emphasis is totally upon discovering how the 'best' outcome for the patient can be achieved.

The role of the dental practitioner is a rapidly changing one. In the past, we were revered for our expertise and we simply informed our patients which treatment they 'should' have. However, it is increasingly recognised that patient satisfaction is enhanced immeasurably where mutual dialogue takes place.

This book does not advocate demand-led dentistry. However, it is based on the surmise that in today's dentistry, it is simply no longer acceptable to be unaware of the consumer voice. The content of this small book describes how this view of decision making is in fact the only way that dentists can continue to make ethical, acceptable and safe decisions, and — importantly — cost-effective decisions.

EJK and NMN
January, 1997

Acknowledgements

The authors would like to thank, firstly, the wonderful general dental practitioners in Glasgow who so generously gave up their time to participate in the research which underpins the words in this book. Secondly, we extend our gratitude to Mrs Sheila Brackin who so painstakingly and cheerfully typed and corrected the manuscript. Finally, our thanks go to Teresa Waddington whose expertise, support, guidance and, above all, enthusiasm allowed this series of papers to become a book.

Contents

1

An introduction to clinical decision making

In order to make the most appropriate decision in a given set of circumstances, an understanding is needed of how people formulate decisions on the basis of their surroundings. Using a case history, this chapter introduces the factors involved in clinical decision making and their relevance in the decision making process. These will then be discussed in depth in later chapters.

Clinical decision making — an art or a science?

Why is it that the choice of treatment for patients with similar problems can vary considerably? For example, if two dentists see the same patient or one patient is seen at different times by the same operator will the same decisions be made? What influences clinical decision making? Although the state of a patient's dentition will drive treatment planning, there are other influences which affect treatment decisions. This book looks at why and how dentists choose particular treatments for their patients and examines the factors that influence the decision making process.

For example, why might a dentist offer one patient a new denture whilst for another, who has very similar problems, he will choose not to do anything? Perhaps the first patient insisted on getting involved in what was happening, while the second was content to let the dentist make all the decisions. As a result, the dentist's decision making process will be different for each patient. Dentists' relationships with their patients, therefore, can influence treatment planning.

Patients' attendance patterns can also affect treatment decisions. For instance, a dentist who was undecided whether to investigate a young patient's tooth where a fissure looked as if it might be developing a cavity, decided to leave it and hoped that the patient would remember to attend his next appointment so that she could keep an eye on the tooth. If the patient had had a poor attendance record, it is likely that the dentist would have been more reluctant to leave the tooth than if she knew that the patient would re-attend.

Likewise, another dentist found it difficult to make a decision about the right treatment for one of his patients who could not understand what was meant by a '50:50 chance' of success with the apicectomy. The probability of a treatment being successful, and the patient's understanding of such concepts, can also influence treatment decisions.

The following case history illustrates the factors that can influence treatment decision making by highlighting the particular points the dentist has to consider before deciding upon the best treatment plan for the patient.

Case history

Sarah is a 27-year-old single woman. She is highly fashion conscious. She was educated at Cheltenham Ladies College, then became a 'hippie' for 4 years. She now owns and runs a riding school and therefore works outdoors. She is underweight, smokes at least 30 cigarettes per day and a diet analysis reveals that each day she drinks at least ten cups of tea, each with two spoonfuls of sugar. She also claims that she has time to eat a 'proper meal' only two or three times a week. During a working day she eats 3–4 chocolate bars, several other sweets and consumes 4–5 tins of sweetened carbonated drinks which she says she needs to give her energy. She wishes to place a 'gold tooth' on the $\underline{|4}$ which will be visible when she smiles. Sarah has no fear of treatment and claims to brush and floss her teeth daily. The results of her oral examination can be seen in fig. 1.1.

Although it is hard to make decisions without actually being able to talk to the patient, the following description includes some of the factors

On examination, it is found that all remaining posterior teeth are heavily restored with:

$\underline{6\|6}$ and $\overline{6\|6}$	extracted
$\underline{\|7\ 4}$	badly broken down but symptom free.
$\underline{\|5}$	evidence of secondary caries
$\underline{5\|}$	missing

Several periodontal pockets of up to 4 mm
Good oral hygiene

Fig. 1.1 Results of the oral examination of the case history patient.

you might take into account before deciding the patient's treatment.

The decision process

Sarah presents a number of problems, and while she has quite a high level of disease, this is not the reason she has attended. The treatment options for this patient are therefore enormous and thus, if several dentists were each to make a treatment plan for this patient, there would be a large variability between them.

Treatment plans might range from offering only preventive advice/ treatment and declining to undertake any restorations until disease progression is halted, to restoring all diseased teeth, providing periodontal surgery and fitting the gold crown as the patient wishes.

One of the factors that could influence the final decision is the patient's involvement in treatment planning. If the patient is a passive or intimidated individual, their participation in the decision making process will be less than if the person were an outgoing, assertive type. These parameters would affect not only the degree of patient involvement in the decision making process but would also influence the dentist's relationship with the patient.

For example, some dentists will follow the wishes of a charismatic and demanding patient, while others would feel it important to communicate the importance of their clinical expertise and persuade the patient that it is disease processes rather than aesthetics which are, and should be, the prime concern.

The patient/dentist relationship is highly dependent on the degree of similarity between the dentist and patient. Perhaps the better choice of dentist for Sarah would be a young female dentist, who would understand Sarah's motivations and desires better than a middle-aged male.

Obviously dentists cannot choose their patients, but it is generally true that dentists tend to practise in middle-class areas, prefer patients whose lives they understand, spend longer with and communicate more with those with whom they identify.

The individuality of the dentist and patient, therefore, can influence the treatment offered within a practice. Treatment criteria are not firm and immovable or solely based on measurement of pathology, but are adjusted according to what is valuable to each patient.

Another problem which Sarah presents is the borderline nature of her periodontal condition. Her pockets do not warrant surgery and yet in a patient of her age, especially someone who exhibits good oral hygiene, such pocket depths are of concern.

Should the alleviation of progressive periodontal disease form the basis of the treatment plan? The answer from dentists to this question would probably be both yes and no. This is because each dentist sets, for each disease and for each patient, a threshold at which they feel treatment is needed. These treatment criteria vary from dentist to dentist and from patient to patient, but an understanding of the perceptual and judgmental factors that affect treatment planning can enhance a dentist's ability to make optimal decisions for each patient.

Before deciding upon a treatment plan for Sarah, a close examination of the patient's preferences is needed.

How important is the visual impact of a visible gold crown to Sarah? Is it more important to her than making her mouth more healthy? If so, how much more important?

On examining the patient's lifestyle, it is clear that she has a very positive attitude to risk. She smokes although she knows it might damage her health because she has a sense of invulnerability: she knows smoking can kill but feels that it will not affect her. She also says that 'no-one can see her lungs'. This gives further confirmation that the patient is motivated towards healthy behaviour by aesthetics rather than evidence of disease.

Dentists' personal risk attitudes, combined with their understanding of their patients' risk attitudes will almost certainly affect the decisions which they make.

Explaining to Sarah how likely it is that she will encounter problems if she does not have some teeth restored is the next problem to be faced.

Although she understands the processes of disease, Sarah feels that she is not personally susceptible to the damage that high sugar intake can cause. Sarah rarely takes action for the sake of her health, and is unlikely to undergo unpleasant experiences in order to achieve a gain some time in the future. Explaining the probability of both good and poor outcomes to patients is difficult and the extent to which a dentist can make patients understand difficult concepts such as 'probabilities' can influence treatment planning and the decisions that are made.

The value which dentists place on filled teeth, or decayed teeth, will clearly be different from Sarah's. A decayed tooth probably has the same value in Sarah's eyes as a filled or sound tooth, so long as disease does not make the tooth's appearance unaesthetic and does not cause pain. Dentists must therefore try to understand and take account of the values which patients place on health outcomes.

Finally, what is to be done about the patient's missing $\underline{5|}$? Sarah, who is clearly very appearance conscious, has not mentioned the gap and yet it is noticeable when she smiles. Should a gold pontic on a bridge be suggested instead of the crown on |4? The final decision about this dilemma, however, might rest on financial matters. Acceptability of treatment to patients is not limited simply to clinical criteria, and this can be a major factor when planning a patient's treatment.

Conclusion

Decision making and treatment planning are very complex processes involving many influential factors, both external and personal to the dentist. These common factors are listed below in fig. 1.2.

This introduction has examined influences on decision making. The

- patient/dentist relationship
  — patient's involvement in treatment planning
  — personality/social similarities between patient and dentist
- patient attendance
- probability of success of treatment
- risk/benefit ratio
  — whether benefits of treatment outweigh the risks
  — patient's and dentist's attitude to risk
- dentist's and patient's values of dental health care
  — are preferences aesthetic or health based?
- dentist's personal treatment threshold
- patient's financial abilities.

Fig. 1.2 Common factors which can affect treatment planning.

following chapters will explore in depth the factors which influence clinical decision making. The aim is to construct a framework that will explain how decisions are formulated on the basis of the individual's environment, applying the principles of decision psychology in a dental context.

Exploring the way in which decisions are made is a fascinating field of study and, although this book could never change the way dentists behave, it may help to explain why and how decisions are made so that the best decision can always be achieved for each patient.

Summary of chapter 1

Factors which can and should influence treatment planning

- **Relationship with patient**
  A patient may be either acquiescent or may demand what he or she wishes. To some extent the dentist must recognise the consumer voice and provide treatment which addresses the patient's wishes. On the other hand, with a patient who casts the dentist in an 'expert' role, it becomes difficult to determine the patient's true personal attitudes and values. Without some understanding of what is of value to the patient, it is almost impossible to make realistic and appropriate decisions. Some participation from the patient is therefore a prerequisite to good treatment planning. It is a mistake to assume that the patient's value system is the same as the dentist's.

- **Patient personality**
  The patient's personality will impinge on both the patient's value system and their attitudes to health, disease and, most particularly, risk. The patient's character will also affect how the dentist and patient interact with each other.

- **Attendance and attributes/values**
  Attendance patterns will, to some extent, reflect patients' dental health attitudes and values. For example, you might assume that if patients attend the dentist for a dental check-up they must be motivated and anxious to maintain their teeth despite the trouble and cost involved.

- **Attendance and dental health behaviours**
  Attendance patterns may also give some indication of patients' dental health behaviours. Patients who attend for regular check-ups and whose value systems make their oral health a high priority are likely to undertake behaviours which are conducive to health, ie they are likely to brush, floss and limit their sugar intake. When planning treatment, therefore, the probability of new disease and the probability of disease progression can be expected to be lower.

- **Attendance and restorative thresholds**
  Regularity of attendance has an important influence on the setting of intervention thresholds. For regularly attending patients with

disease which is on the borderline of needing treatment the dentist knows that she is likely to see the patient before problems arise. Therefore, leaving disease untreated will not cause any untoward occurrences. On the other hand because seeking regular check-ups indicates positive dental attitudes the patient's expectations of dentistry may be higher. This may also influence the dentist's treatment threshold.

- **Costs and consequences of treatment**
For each individual patient, dental treatment and dental disease have a 'cost' attached to them. This does not only mean financial costs but also costs in the sense of psychological, emotional or social burdens (for example, who will look after the children while mum is having treatment?). The 'costs' of disease often only become apparent to the patient when it is too late—pain has arisen. Patients will only take action if they view the consequences of their action as being of greater value than the cost. For some people, the knowledge that their teeth are in perfect health is worth all the effort required in order to make them so. Thus, a main consideration when planning treatment would be whether the benefits for the patient (the gain in health and well-being) are worth the expenditure (in money, time, angst, anxiety, discomfort).

- **Patient age and treatment planning**
The patient's age affects both the probability of disease (90% of new caries occurs prior to the age of 15) and the likelihood and speed of disease progression. Similarly, age has an important effect on patients' behaviour, with younger patients mostly adhering to their parents' routines, teenagers 'doing their own thing', young single adults increasingly becoming sexually aware, and adults adopting the routines which fit most closely with their local culture and daily lifestyles. Each of these considerations could and should influence which treatment plan suits an individual patient. In particular the type of advice offered, and the format in which it is presented must be tailored according to the patient's age.

- **Patient's diet and treatment planning**
The patient's dietary patterns will affect the probability of caries and speed of caries progression. It may also reflect the patient's attitudes and values towards health and healthy behaviour, and may reveal their attitude to risks and benefits.

- **Patient's attitude to risk and treatment planning**
  A patient's attitude to risk will profoundly affect the treatment plan made for them. For example, some mothers would far prefer their child to have teeth taken out under general anaesthetic (even though general anaesthetics are associated with a very small risk of death) than face the very high probability of their child finding a local anaesthetic extraction distressing. Mothers seem, therefore, to be more willing to take very small risks of a terrible outcome rather than a high risk of a mildly unpleasant outcome.

- **Patient's lifestyle and treatment planning**
  Examination of the patient's current lifestyle may reveal some of their attitudes to risk, and will therefore give an indication of what is likely to be the optimum type of treatment plan.

- **Patient's occupation and treatment planning**
  Patient occupation may be a key to practical treatment planning. Firstly, their occupation may dictate to some extent their dental needs, for example do they perform in public at all? Singing, public speaking, playing a musical instrument will all affect the patient's dental needs, both in terms of the functional requirements and the aesthetic requirements.
  Secondly, a patient's occupation will dictate the amount of time and money which they have to spare on dental treatment. Closely supervised factory workers are less likely to be able to take paid time off work for dental treatment than professional office workers.

Case history
The following patient presents for treatment. Perhaps you would like to reflect on the questions you would need to ask and the information you would need before making a definitive treatment plan.

> Mr O'Neill is a 65-year-old man who has just retired from a lifetime in farming. He plans to emigrate in the near future. He has not attended a dentist for approximately 15 years.
> Intra-orally, he has 19 standing teeth, none with restorations. All his molar teeth are missing, as is the ⌊2. His oral hygiene is reasonable although he has 3–4 mm recession on all upper teeth.

Case history — our suggestions
We would wish to question the patient regarding the reason for his sudden attendance. If he were symptomless, we would suppose that the prompt to the visit may be social (his daughter is getting married and he wishes the |2 replaced) or perhaps financial (he may be concerned about the mechanisms of acquiring dental care once he has emigrated) or perhaps he wishes to take advantage of the last NHS dentistry available to him. It is of course also possible that Mr O'Neill simply has more time available for activities such as attending the dentist.

Therefore, we would:

(1) ascertain whether the patient perceives any problems with his teeth
(2) ascertain whether the patient feels any necessity to improve the state of his dentition
(3) ascertain the time, money and effort the patient is prepared to commit to achieving any improvement.

For example, if attendance at a wedding *were* the motivation to attend in order to achieve an improvement in appearance, or if the patient simply wishes an 'overhaul' before emigrating, it would be inappropriate to offer Mr O'Neill a chrome cobalt denture with precision attachment stress breakers for the free-end saddles — even if such an appliance is considered to be the last word in dental prosthetics. This is because:

(a) it is inappropriate to Mr O'Neill's needs and therefore the cost would be highly likely to outweigh the benefit
(b) it would be expensive and difficult to maintain
(c) a number of lengthy procedures would be needed before the denture would be ready to be worn.

If the reasons for Mr O'Neill's attendance were social/aesthetic, perhaps a better solution would be a simple Maryland temporary bridge to replace |2 (in time for the wedding if necessary). This would require minimal preparation, would not require a high level of maintenance, could be inexpensive and would address the patient's needs.

The best treatment for a patient is not necessarily the most complex and costly one. However, careful questioning in relation to all the points included in the summary should ensure that a suitable treatment plan is negotiated.

Points you may wish to consider

- what other factors should influence treatment planning besides clinical features and levels of disease?
- which is more important in determining the optimum treatment plan? Patient's age, diet, risk attitude, lifestyle or occupation — or are they the same?

2

Making sense of treatment decisions

Clinical decisions made by dentists can vary considerably. Although there has been a tendency in the media to attribute these differences between practitioners to deliberate unethical practice, as chapter 1 concluded, variations in decision making are a result of the complexity of assessing the risks of different treatment options and evaluating the outcomes of treatment. This chapter looks at why dentists can make different decisions when faced with identical cases (variation), why sometimes the 'correct' decision is not made (error), and also looks at the other issues which affect treatment decision making once disease has been perceived.

Introduction

As Chapter 1 demonstrated, variation in decision making can be the result of the complex process of choosing the best treatment option within the given circumstances. As well as the need to decide upon the best treatment for a particular kind of disease, other issues such as patient and environmental factors are vitally important. Essentially, however, differences between treatment decisions can be considered to stem from two main sources: *perceptual variation* and *judgemental variation.*

Sources of variation

Figure 2.1 is a colour chart showing two people's response when asked to decide when a shade of blue can no longer be defined as 'blue'. The two responses are very different. These differences are the result of variations between the two people's perceptual and judgemental qualities.

Perceptual variation is when people 'perceive' things differently. For example, if two people are judging a colour they might perceive it differently because one might have a form of colour blindness, or one might view the colour patch using a coloured light source. As a result both 'see' different colours. A practical example of this would be when one dentist sees a white spot lesion but another does not, perhaps because of poor lighting.

People may also have different opinions. For example, two people might disagree about when the colour 'blue' becomes so dark that it ought not to be called blue. This is judgemental variation. It is possible to predict this type of variation, for example in fig. 2.1 the two opinions are more likely to agree at the extreme ends of the colour shading but disagree about colours in the middle of the range. Judgemental variation is a result of differing values about what constitutes a 'positive case'.

For example, one dentist might consider that caries visible on a radiograph warrants restoration, whilst another dentist might insist that the caries must appear to penetrate the ADJ before restoration is justified.

Perceptual variation
Occurs, for example, when dentists examining the same tooth site disagree about what they are looking at: they 'see' different conditions. These dentists' treatment decisions can differ because they are seeing different levels of disease.

Judgemental variation
Occurs, for example, when dentists examine a tooth and agree about what they see, but still disagree about how the condition should be treated. The dentists have different opinions about the appropriateness of a treatment; it stems from their judgement about what is the right thing to do under a given set of circumstances.

It is important to note that differences between treatment decisions stem from two very different sources because if all dentists were expected to offer patients exactly the same treatment, any variations would have to be controlled. The means of controlling these differences would differ according to whether it was perceptual or judgemental variations causing the differences.

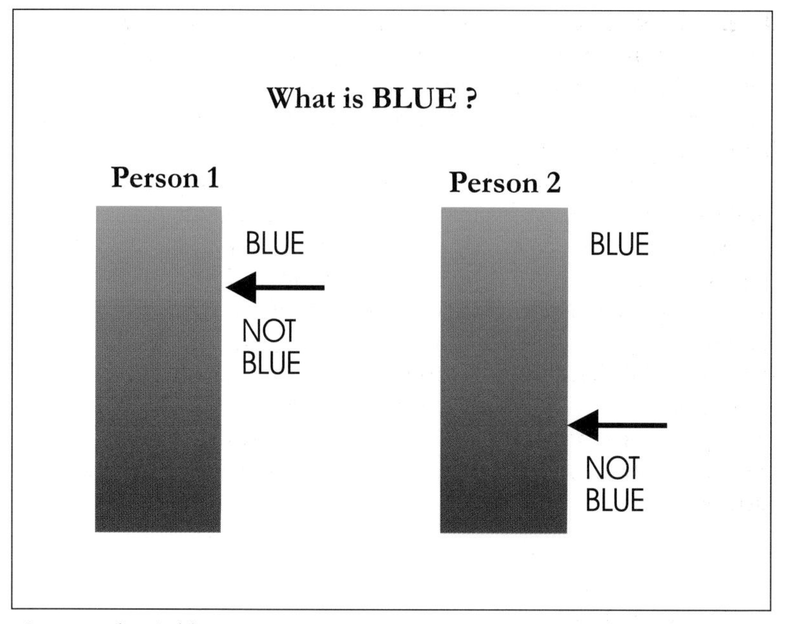

Fig. 2.1 What is blue?

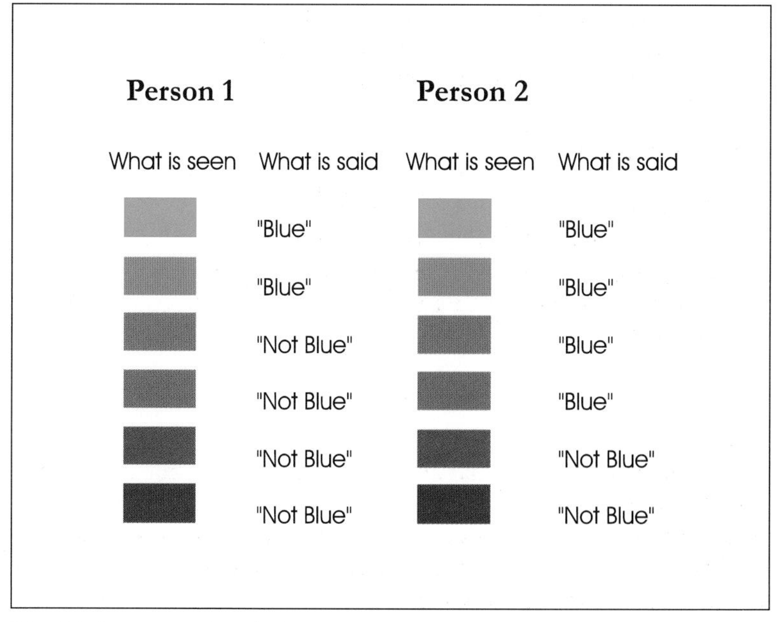

Fig. 2.2 Judgemental variation

Finding ways of reducing the likelihood of perceptual variation is much more difficult than reducing judgemental variation. Perceptual variation is random in nature and therefore it is difficult to predict when it is likely to occur. However, perceptual differences may also be a less persistent form of variation than judgemental, in that many disagreements are just one-off events. Perceptual differences, therefore, can probably be substantially eliminated by double checking or reviewing cases.

In contrast, judgemental variation is a predictable process. Using the blue colour chart example again, it is possible to predict whether individuals will agree or disagree about the colour of a particular patch in fig. 2.2 by taking into account the point at which each person classifies the colour 'blue' as blue in fig. 2.1.

Because of its predictability, judgemental variation is modifiable. Specific criteria can be established to reduce the decision variation margins, for example by defining the colour blue by wavelength. Decision aids could be constructed such as colour matching charts, and judges could undergo specific training to help them use the criteria and decision aids reliably. Similarly, dentists can be persuaded to make similar judgements about a given case by establishing strict criteria guidelines, for example deciding whether a given level of disease warrants treatment or not.

Why do dentists vary?

Perception is an active process. Every individual places a particular meaning on what they see according to what they remember of their past experiences. For example, if a dentist was sued for negligence for 'missing' a carious cavity, he might become 'hyper-perceptive' of anything which might possibly indicate the presence of caries.

Although the above is an extreme example of how experience alters perception, it is important to remember that each individual dentist's experiences will be different. For instance, the discovery of a large occult carious lesion in a regularly attending patient is likely to affect how the dentist perceives teeth of similar appearance in the future.

A dentist may, in response to past findings, develop a very complex set of cues and clues which indicate, even if only at a subconscious level, that disease is present. Many clinicians, for example, are perhaps unaware of the extent to which they are discriminating between minute changes in colour and shadow when they are examining the fissures of teeth.

Judgemental variation affects what dentists decide to do about disease once it has been established that there is a case of disease. As well as being influenced by patient and environmental factors, a dentist's judgement is affected by personal factors such as the dentist's individual treatment threshold and attitudes to risk.

Issues involved in the decision making process

It is important to recognise that what dentists decide to do once they have 'seen' or 'perceived' a sign of disease is dependent both on their own judgement and on patient and environmental factors.

Figure 2.3 is a model of the decision making process. The dentist is the central character in the decision making but is not isolated from environmental or patient factors. The double-headed arrows illustrate that there is a reverse flow between decisions made and the factors influencing them.

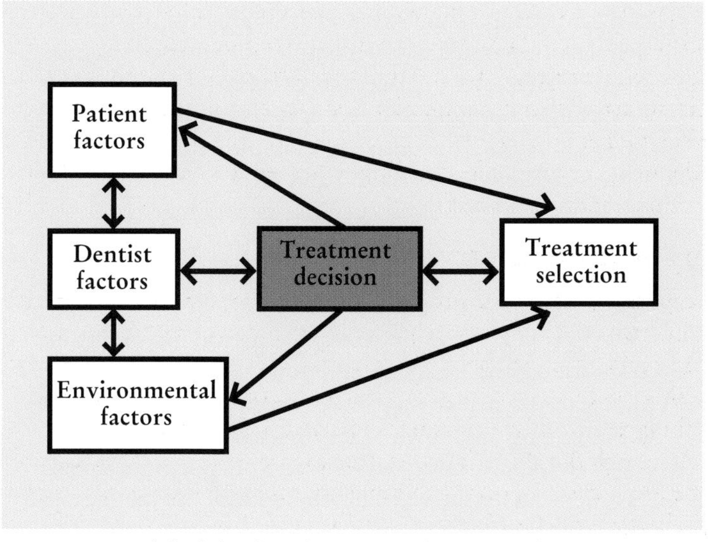

Fig. 2.3 A model of the dental treatment decision making process

Patient factors

Often a patient has already decided that something is wrong before they visit the dentist. Under these circumstances part of the dentist's task is to detect the cause of an identified problem rather than discover problems the patient is unaware of.

Once a decision has been made that a person requires treatment, the role of the person involved changes from being a person being screened for disease to a patient being prepared for restorative dentistry. In the latter role, various aspects of the person may change: he or she may for instance become more willing to have additional work done once it has been accepted that a filling is needed.

The extent to which patient factors may affect a particular dentist is likely to vary in relation to the dentist's views about the 'ownership' of the patient's oral health: how much does the dentist act as the custodian

Variation or error?

Treatment variation between dentists is not synonymous with error because variation is a result of perceptual and judgemental differences which in turn can be influenced by patient and environmental factors. Because these factors can change from case to case, variation is a natural and positive result of the complex process of decision making. Error, or an inaccurate decision, is not the result of differing opinions and perceptions but is an objective miscalculation. The following example demonstrates the difference between variation and error.

In the eighteenth century the astronomer royal was plotting the course of a planet. As this tended to be a particularly cold and tedious job he alternated his nightly shifts with an assistant.

After taking readings over a period of nights he transcribed the readings and found something rather unusual: the planet they were studying appeared to be wobbling up and down as it crossed the sky.

Newtonian physics was spared from potential embarrassment however, when closer scrutiny of the findings showed that if just one of the astronomer's results were studied the orbital path looked fairly normal.

They concluded that the unusual orbit of the planet was due to a systematic error between the way the two astronomers took their readings. At least one (possibly both) of the astronomers must have been wrong from some objective viewpoint, if such a viewpoint could be established.

of a patient's oral condition as opposed to acting like an adviser to a client? Does the dentist allow the patient to become part of the treatment decisions once the significance of what has been perceived has been described and explained? Is the patient only involved in a passive sense in so far as the dentist makes an evaluation of the best course of action on behalf of the patient and simply informs the patient about what this will be?

Environmental factors and treatment thresholds

The immediate environment is likely to change once a decision to treat has been made; instruments will begin to be prepared and this in turn may affect the dentist. Once a decision to fill one tooth has been made, the perceptual process or the judgemental decision threshold that a dentist had used previously might be changed.

For example, a decision that caries in need of treatment is present in one tooth may increase the dentist's expectation of finding additional lesions in other teeth, or may reduce the stringency of the criteria used to decide whether to fill other cases of caries in the same mouth. It might be preferable to 'tidy up' any borderline cases of treatment need once some sort of treatment has been decided upon, rather than leave them until the next course of treatment.

Range of treatments and risk factors

The range of treatments available for management of a condition also affects whether or not a decision to treat is made. If a condition is very serious and a treatment option exists with few side effects and a high rate of success, then it is easier to make a positive decision as the consequences of a false positive decision are not too traumatic to the patient.

However, if the disease is fairly innocuous and the treatment can have unpleasant side effects, then decisions to treat may be limited only to cases where there is absolutely no possible alternative: the dentist will err on the side of caution in order to ensure that no false positive decisions are made.

Dentists have their own perceptions, treatment thresholds and attitudes to risk whereby they decide upon their patients' disease management. These treatment criteria can vary considerably between individual dentists and are adjusted according to what is important for each case. Research about treatment criteria reported to be used by general dental practitioners suggests that general factors such as age, practice location and place of training seem to have very little overall effect on dentists' views about when a carious lesion ought to be treated.

This suggests that factors unique to each individual dentist may play a much greater role in dental treatment decision making. These individual factors and their effect on treatment planning will be explored in future chapters.

Conclusion

The reasons why treatment variation occurs are wide ranging and complex, involving patient and environmental factors as well personal perceptual and judgemental factors. Figure 2.4 summarises the issues involved.The sheer complexity of all the inter-relating factors which come into play when a treatment decision is made, illustrates how

- treatment variation stems from two separate sources — perceptual variation and judgemental variation.
- perceptual variation is when people 'see' things differently. Why dentists perceive the same condition differently may be due to past experiences or to environmental factors such as poor lighting.
- judgemental variation is when people regard the same condition differently and so decide on different treatment options. Why dentists judge the same condition differently may be due to patient and environmental factors or to treatment thresholds, attitudes to risk and past experiences.
- the terms variation and error are not synonymous.

Fig. 2.4 Issues underlying treatment variation

difficult it is to attempt to rationalise the decision process. Nevertheless, it is hoped that this introduction to the decision making process will lead dentists to examine how they come to choose the most appropriate treatments for their patients.

This book may not influence individual dentists' treatment planning, but because the profession must be able to defend its own judgements if it is to retain autonomy, it is hoped that it will offer explanations about decision reasoning which will enable dentists to prevent themselves and the dental profession from being branded as inaccurate, unethical or unscientific.

Summary of chapter 2

Factors influencing variation in treatment planning

- **Judgemental variation**
  Judgemental variation occurs when dentists agree about what is seen, but have different views about what treatment is needed. For example, one dentist might believe that an incomplete root filling indicates a need for re-root treatment, whilst another might believe that signs of inflammation are required before such treatment is warranted. It may be controlled by the adoption of standardised clinical guidelines.

- **Perceptual variation**
  Perceptual variation (seeing the case differently from a colleague) is random in nature and therefore it is difficult to predict when it will occur. It may, however, be controlled by re-evaluation of the case at a later date, using some other diagnostic system to confirm diagnosis or seeking a second opinion.

- **Effects of variation**
  Dentists may make different treatment plans for the same patient because of perceptual and judgemental differences. These in turn are influenced by patient and environmental factors.

- **Variation and treatment planning**
  The patient (including the clinical factors) may be perceived differently by two dentists. Their interpretation of the observations may then influence their judgement, ie the decision about what would be best for the patient. If dentists perceive different levels of disease, this may then alter their perceptions of the patient as a person. The dentists are therefore likely to choose different treatment strategies.

- **Deciding who is 'right' and 'wrong'**
  If, for example, two dentists make a different treatment plan for one patient it is not possible to determine who is 'wrong' overall. It may be possible to check the dentists' perceptions by reviewing cases etc., but each judgemental decision must remain invalidated. Formal decision analysis (decision trees/sensitivity analysis) may, however, identify the most appropriate action to take for a given dental condition.

A simple exercise
Perhaps you would like to try this exercise with your colleagues.

(1) Find a suitable bitewing radiograph. Which tooth surfaces would you restore?

(2) Compare your treatment plan with a colleague. If possible, make a consensus treatment plan, then compare this with one made by another colleague. Discuss why there are discrepancies between the various plans.

- For example, did each observer see the same number of lesions?

- Did each observer agree about the depth of the radiographic lesion?

- Did the observers agree about the depth of the radiographic lesion but disagree about how radiographic appearance translates to clinical lesion depth?

- If all observers agreed about the presence and depth of lesions, did they make similar treatment plans?

As discussed earlier, there is no single 'right' answer to this question, but your discussions should have been instructive.

If the observers with whom you discussed the radiograph counted different numbers of lesions, the discrepancy arose from perceptual variation. Similarly, if the observers disagreed about the radiographic depth of the lesion, these differences again arise from differing perceptions.

In contrast, differences in opinion about the relationship between radiographic lesion depth and actual clinical lesion depth stem from judgemental variation. Finally, opinions about how to treat a lesion of a given depth may vary. This, again, arises from judgemental variation.

Points you may wish to consider
- What is meant by the terms 'perceptual variation' and 'judgemental variation'?
- Perhaps you can describe a dental example of each type of variation.
- Do you consider it more important to control perceptual and judgemental variation? Can you make suggestions as to how this might be done?

3

To treat or not to treat?

Chapter 2 examined the two principle sources of variation: variation stemming from perceptual differences such as diagnostic errors, and variation stemming from judgemental differences such as different treatment strategies. This chapter looks at judgemental variation in more detail and considers the nature of practice variation. It asks whether such variations are a sign of good or bad clinical practice.

Introduction

Practice variation is a ubiquitous phenomenon. It has been noted in tonsillectomy rates, admission for back pain, surgical procedures and Caesarean section rates. Variation in practice between dentists is often discussed as if it were synonymous with malpractice, but is this a fair deduction?

Academic studies have shown that variation among dentists occurs even when the clinicians know they are being compared with others and have no financial incentives for planning a particular treatment. This suggests that there is a great deal of variation between dentists as professional decision makers and that it is 'natural' rather than unethical.

However, observing variation in practice raises questions about the quality and appropriateness of the treatments being prescribed. Such questions in turn raise issues concerning the cost of services which seem to be provided in an arbitrary way.

Practice profiles and patient preferences

Some would argue that if all of the procedures undertaken were necessary, they would be carried out at an equal rate in all practices and areas, and all dentists would earn similar amounts of money and do similar treatments for their patients. However, two different schools of thought present explanations about why one practice does not supply its patients with the same type of care as another. One suggests that practices vary in the treatments they provide because patients attending one practice are different to patients obtaining care from another. Alternatively, the other school of thought believes that the practice styles of dentists may differ.

The second explanation suggests that some uncertainty exists about what constitutes the optimal choice of treatment: differences between dental practices suggest there is no consensus view about what treatments produce the most effective type of care.

The first school of thought, however, implies that practice variation may be due to real differences in patients' needs. The severity or incidence of disease is the factor which causes different practices to provide different types of care. This explanation implies that practitioners do know how to make optimal decisions and that the observed variations in the treatments provided are due to differences in the levels of disease among patients, or that different types of patients attend different types of practices. It therefore appears that variation between practices is perceived to be good if it is related to patient characteristics, but bad if it is caused by other factors.

However, this view of good and bad types of variation is too simplistic. The previous chapters have attempted to demonstrate that variation is a result of the complex relationship between oral pathology and treatment need. But why are decisions about when to treat and how to treat so complex, and are the variations that result desirable or undesirable?

Treatment criteria

It is sometimes convenient to think that a decision to treat is made when disease passes a certain point at which leaving it alone is no longer an option. The term for this point is the *treatment threshold* or *treatment criterion*. However, the treatment threshold concept of when to treat/not treat, does not sufficiently explain why clinicians choose a particular decision and why these decisions can vary considerably from case to case.

Many clinical cases have a grey area where a range of different treatment plans would be suitable. Consequently, treatment thresholds used by dentists are not rigid and definable points separating 'cases in need of treatment' from 'cases which do not need treatment'. Rather, treatment thresholds may be fairly broad sets of standards indicating types of cases for which it is easy and simple to make a treatment decision, and cases where many different options, including inaction, may be considered.

As a result of this grey area, there are opportunities for error when deciding upon the best treatment option. Figure 3.1 shows how the choice of a threshold can affect the chances of making a wrong treatment decision. Within this grey area there will be some cases which will benefit from treatment and some which will not, but there is no ethical fail-safe way of discriminating between the two.

Fig. 3.1 shows what happens as the treatment criteria becomes more, or less, stringent. For example, using point A as a decision criterion to decide whether or not to restore a questionably carious tooth, would mean that very few teeth would be unnecessarily filled. No sound teeth would be treated (false positive decisions) by choosing such a threshold, but in order to achieve this, a number of cases which do require treatment would be left untreated (false negative decisions).

At the other extreme, using point C as a decision criterion would ensure that all carious teeth would be filled, but using such a threshold would mean that many teeth which are not carious would also be filled.

A criterion somewhere in the grey area will lead to errors of both sorts. Therefore, when a dentist makes a treatment decision for a patient,

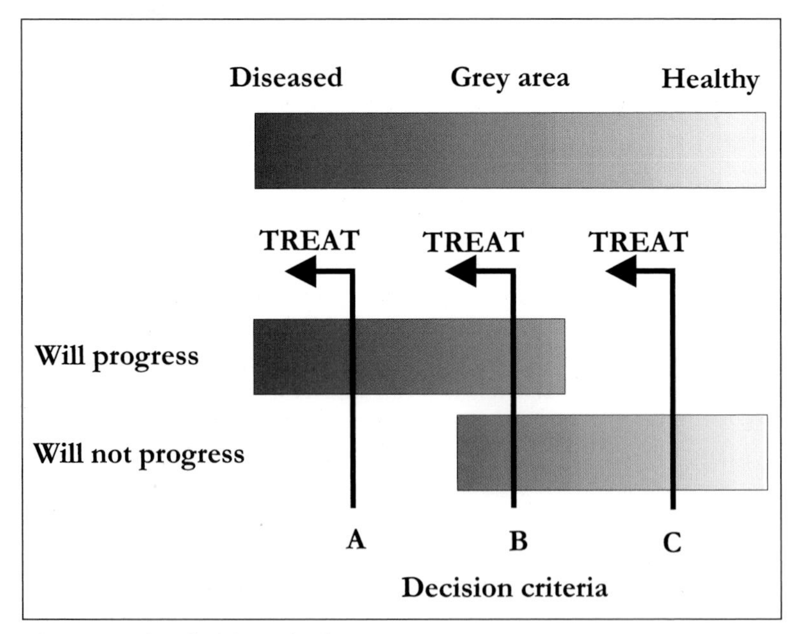

Fig. 3.1 Setting decision criteria

that dentist is really deciding which type of potential error is most likely and, at the same time, judging which type of error is least harmful. This suggests that treatment variation is not a sign of bad practice, but a sign of good patient care because the dentist must assess each case individually before deciding on a treatment plan.

Quality care

Achieving quality in dentistry is a growing concern amongst the profession and the prevailing view, particularly in the media, that quality implies uniformity of treatment is particularly damaging to the profession's efforts to improve the quality of patient care.

The argument that a given type and extent of disease must warrant a particular type of treatment suggests that treatment is dictated purely by disease levels. This opinion is flawed in many ways including the fact that it ignores the 'grey' areas in cases where the treatment/no treatment options offer different sorts of advantages to the patient. Variation between treatments planned for similar conditions is acceptable and is

- academic studies suggest that treatment variation is natural and not a sign of malpractice

- whether to treat or not and how to treat involves a complex decision making process because:
  - dentists must arrive at the best solution for each particular patient and assess the value attached by the patient on the outcome
  - many possible treatments are often suitable for the one problem — the 'grey area'
  - the dentist must assess the potential risks from each treatment option

- the choice of treatment threshold affects the types of errors made — false positive or false negative decisions

- as dentists assess each case on its own merits, treatment variation can result from high quality patient care

Fig. 3.2 Treatment variation

indicative of quality treatment planning, providing there is a rational basis for the choices that have been made.

For example, extracting a decayed and painful tooth offers the patient immediate relief from pain and a certainty that the pain will not recur. The alternative treatment (restoring the tooth) may relieve the patient's pain, but there is a risk that the problem might recur. The small risk of further trouble from the tooth if it is restored rather than extracted is weighed against the problems of losing a tooth, as well as other consequences which such a course of action might entail (will it need a prosthetic replacement for example).

Quality care, therefore, cannot rest simply upon making decisions according to the disease and disease levels. In essence, good quality decisions entail choosing a course of action that will increase the probability of an outcome which the patient regards as favourable, while at the same time reducing the probability of an outcome which the patient will regard as unfavourable.

Advantages of restoring tooth	*Advantages of leaving a tooth unrestored*
Caries progression stopped	Tooth remains completely healthy if no caries is present
Pain unlikely	You have avoided unnecessarily removing tooth tissue
Patient experiences restorative care before symptoms arise	The patient is happy
Certainty that first permanent molars are not diseased	You will have an opportunity to observe progress (if any) at next visit
The caries may be deeper than it appears and thus potential pulpal involvement will have been averted.	Patient does not have to have restorative treatment
	The decision is reversible. Even if caries is present, its progression is slow enough for you to observe before opting to restore in a patient who goes for regular checkups.

The value a person places on the outcome will vary according to the attitudes, behaviour and preferences of the patient. For example, returning to the extraction/restoration problem, the decision will be influenced by how much the patient wishes to retain a tooth, their attitude to oral health in general and restorative treatment in particular, and by past dental experience. This is entirely different from the belief that a given disease always requires a given type of treatment.

Quality care, therefore, is provided by a dentist who chooses the optimal treatment for each patient, taking account of their lifestyles, needs, attitudes and wishes. To make such optimal decisions, the dentist must accurately assess the risk or probability of poor and good outcomes and determine the worth or value attached by the patient to the resulting oral health state.

Conclusion

Treatment decision making is not merely a case of deciding whether or not to treat disease. There are many factors to be considered for each case before the best treatment option can be chosen. This chapter has aimed to prove that variation in treatment decision making is natural and good — a sign of quality patient care, and not, as some would suggest, a sign of bad practice. Fig. 3.2 summarises these points.

Summary of chapter 3

Factors influencing practice profiles

- There are two main explanations of practice variation. One theory is that the types of patients (and hence the type/pattern of disease) visiting one practice are different from the types of patients visiting another. Therefore, patients' needs (in physiological, psychosocial and pathological terms) differ between practices. If the dentists working therein are to adequately meet their patients' needs, they must offer different types of treatment.

- The second common explanation of practice variation suggests it is simply that dentists choose to behave differently. This explanation implies either that there is a 'best' or 'correct' treatment for each type of disease, but that not all dentists manage to make that optimal choice, or alternatively, that no-one actually knows what is the right or best treatment and that the choices made by dentists are fairly arbitrary.
- Although both theories have some truth, neither completely or satisfactorily explains why practice profiles differ as widely as they do. Although practices may attract dissimilar patient pools, it is well known that dentists' practice profiles and treatment strategies can differ even when faced with very similar patients. These variations stem from the fact that the relationship between treatment need and the amount of disease present is not a simple linear one, but is influenced by many external factors.

- The treatment threshold is the 'point' in the progression of a disease at which inaction/non-treatment is perceived by the clinician as being likely to lead to a poor outcome for the patient. However, because clinical cases have 'grey areas' and because of the many external factors which impinge on decisions, it is not possible to define an exact 'point' when treatment is, or is not, necessary. Therefore, although the terms treatment threshold or treatment criterion are commonly used, in actual fact the 'thresholds' used by clinicians need not be rigid and definite points.

- Some dentists will set treatment thresholds at a level which ensures they never intervene except when there is no other possible course of action. Others will treat patients who they think 'have a good chance' of benefiting. Finally, some clinicians opt to treat all cases except those where there is absolute and incontrovertible evidence that no disease or pathology exists. Furthermore, one dentist may operate at either extreme, depending on the type of patient he is dealing with.

- Quality care involves making decisions which will increase the probability of an outcome or process which the patient regards as favourable while, at the same time, reducing the probability of an outcome or process which the patient regards as unfavourable. Whether the outcome is regarded as favourable or unfavourable will vary according to the attitudes, behaviour and preferences of the patient, ie quality is about meeting the patient's requirements.

Case history

The following patient of yours offers a real dilemma.

Peter is a 15-year-old schoolboy who is a regular dental attender and has excellent oral hygiene. At a routine visit, you notice suspicious shadows beneath the occlusal surfaces of the lower first permanent molars. Clinically, the fissures appear deep and stained, but otherwise normal. You cannot perceive any signs of cavitation.

Perhaps you would like to mentally list the 'pros and cons' of offering restorative treatment to this patient.

Case history — our suggestions

This dilemma represents the grey area of treatment planning. It is impossible to know for definite the extent and rate of progression of any caries in Peter's teeth since radiographs are unreliable for predicting the extent of occlusal caries. The treatment decision therefore rests upon whether a false positive decision (an unnecessary restoration) represents a better or worse outcome than a false negative decision.

Points you may wish to consider

- Dental practices sometimes offer completely different types of treatment to their patients, even though the patients are very similar. Why do you think this occurs?
- Terms such as 'quality care' and 'treatment thresholds' are becoming more and more common. What do you understand by these phrases?

4

Assessing risks and probabilities

Deciding whether a patient requires any dental treatment will depend on an assessment of the risks involved. How likely is it that the condition will get worse if it is left untreated? How likely is it that a particular treatment will be successful? This chapter looks at risk assessment and examines how the relative risks of a particular course of action might be communicated. Dentists should be able to explain clearly and concisely to patients what is likely to happen.

Introduction

The task of the dentist is to choose the optimal treatment for each patient. This requires a rational assessment of the risks involved in a positive or negative decision, and an assessment of the value attached by the patient to the resulting dental health state. To make the best treatment option, dentists should determine their patients' understanding and evaluation of both the risks involved and the resulting dental health state. By involving patients in the treatment planning in this way, dentists are helping their patients make an informed choice about their own treatment preferences and, as this chapter concludes, this will inevitably contribute to better dentist/patient relationships and improved patient satisfaction.

Assessing risk

People are natural assessors of risk. Indeed one theory suggests that people actually crave a certain level of psychological arousal which comes from taking risks. The theory argues that wearing seat belts encourages faster driving, not because drivers feel safer, but to compensate for the decreased level of psychological arousal as a result of being at less risk when a belt is worn.

This view is contentious, but people's behaviour in relation to road safety is a good example of how bad people can be at accurately assessing risk. For example, occasionally people do not fasten their seat belts when driving. Their perception may be that making very short journeys or travelling along quiet or private roads is fairly safe and so decide not to use a seat belt. Their assessment of a comparatively low risk in such cases may be correct. But if given the opportunity to drive a Formula One car on a racing circuit most people would probably want to wear fireproof clothes, a helmet and seat belts, despite the fact that there would be no oncoming traffic and that modern Formula One cars seem to be able to survive extremely high speed impacts under which an average saloon car would crumple.

A dentist's assessment of the risks involved in a particular treatment strategy will be influenced in part by whether he or she has experienced the event before. For example, if every tooth with radiographic evidence of caries is investigated and found to have deep decay, the dentist will come to believe that radiographic evidence of decay is highly indicative of advanced disease.

One way that a dentist can assess risk is to estimate the likelihood or probability of bad and good outcomes. This can be done either by systematically reviewing how often an action has led to a particular result or, as is more often the case, by estimating the likelihood of the outcome in an implicit way, by relying on memory. Unfortunately, the latter way of estimating probabilities is often likely to be inaccurate because memory is not a reliable source of information.

According to psychologists, people tend to remember surprising and

Assessing risk from memory

A dental example of the less reliable way of assessing risk by memory is the 'occult' caries problem. After seeing and investigating many suspicious-looking fissures, and having discovered only minimal lesions you may on one occasion encounter a large and unexpected carious cavity beneath a seemingly sound fissure. This is likely to remain in the forefront of your mind and lead to a tendency to over-estimate the prevalence of this type of lesion. A consequence of this over-estimation might be that you will alter the rate at which you investigate suspicious fissures.

distressing events clearly and, as a result, may tend to over-estimate the probability of them reoccurring. For example, imagine you have recently root treated a lower incisor. Three days after completion of the treatment, the patient returns with pain and swelling. You find this an unusual and surprising event. You have previously undertaken 99 similar treatments which have been successful. When the next lower incisor presents for root treatment, would you accurately state the probability of successful treatment to be 99 per cent? The answer to this question is probably not.

Another problem is that people tend to estimate probabilities in a way that enhances their image of themselves and as a result are likely to over-estimate their own success rates. It has been shown, for example, that junior doctors in acute medical wards can accurately estimate overall patient survival rates, but tend to over-estimate the survival rates of their own patients. This understandable difficulty in accurately assessing probabilities is termed 'ego bias'.

Thus, the most accurate assessment of the probability of a particular outcome subsequent to a clinical intervention may be made by assimilating both the results of scientific investigations, and a systematic review of one's own personal rates of success/failure.

Ego bias
A dentist may have learnt from the literature that 50% of dentists' amalgams fail within 8 years. However, the same dentist may not accept that this applies to fillings he has placed. This may lead to the dentist being less likely to err on the side of caution when making restorative decisions than a dentist who recognises a high failure rate.

Communicating probabilities

Having assessed risks and probabilities as accurately as possible, it is important that they are clearly communicated to patients so they can decide what risks they are prepared to take, and make an informed choice between the treatment options. Before explaining the risks to a patient, it is important that the dentist knows how to explain them in a way the patient can understand. This is difficult to determine and requires good communication skills from the dentist, because people understand and interpret information differently.

For example, imagine you have identified a condition which under normal circumstances would leave you with no alternative but to extract the tooth. However, four new treatment options have just been developed for this sort of case. Look at figure 4.1 and before reading on, quickly choose which of the treatments you would prefer.

The exercise shows that people's reaction to numerical data may vary. These differences between people's reaction to data could lead to misunderstandings, for example when a dentist tries to explain the

Scenario:
There is a new dental condition for which the only current treatment is extraction; four alternative new techniques have been developed. Which would you choose?

1 this technique will only save 60% of teeth

2 40% of teeth treated by this technique will need to be extracted

3 this technique will save 3 in every 5 teeth

4 400 in every 1000 teeth treated by this technique will need to be extracted

Which choice of treatment did you choose?
If you did not spend too much time working out that all of the options in fact yield the same result, you may have chosen a phrase which had a 'positive' frame: it emphasised the good side of the outcome, as in treatment 1 or treatment 3.

Alternatively, you could have made a particular choice because you are more comfortable with smaller real numbers such as in the phrase '3 in every 5'.

Or you might be more impressed with larger real numbers where the probability was expressed as a proportion out of 1000.

You may also have been influenced by value judgements which can slip into phrases such as the use of the term 'only' in treatment 1.

Fig. 4.1. Which treatment would you choose?

possible risks of treatment to a patient. This implies that it might be better to use phrases in preference to numbers to describe the chance of something happening, such as 'poor chance', 'doubtful', 'perhaps', 'reasonable to assume', 'likely' and so on. But these too may not give patients a realistic idea of the true probability of occurrence of an event.

Research has shown that doctors and patients may have very different interpretations of such phrases. Most doctors seem to regard the concept of a 'small risk' as 1 in 1000 at most, and often much smaller. A doctor might be horrified at the idea of undertaking treatment which has a 1% risk of death for a non life-threatening condition. Yet many patients, when asked to translate the types of phrases listed above, equate a probability of 1% with a phrase like a 'small risk'.

One factor contributing to this is likely to be the difference in perspective between the practitioner and patient. A doctor who treats 100 similar cases each year would probably see one death per year resulting from this treatment of an otherwise non-fatal disease as unacceptable. Many patients, however, are only faced with the issue on a single occasion, and would be prepared to take what to them, as an individual, may seem to be a fairly small risk of dying.

A further drawback of non-numerical phrases for expressing probabilities is that they will inevitably get confounded with the 'value' of the outcome. A 5% risk of losing a tooth, for example, might be a 'small risk' to a patient whereas a 5% risk of death may be regarded as fairly high.

Therefore, before meaningful communication about risks and probabilities can take place, the dentist must make a careful assessment of the patient's evaluation of the outcome and how best to convey the information to the patient concerned.

Risk acceptance

Risk acceptance is the level of risk a person is prepared to accept in order to achieve something. People's risk acceptances may vary considerably because risk positivity is determined by each individual's understanding of the risks involved and the value placed on the resulting outcome.

Because a patient's level of risk acceptance will affect the choice of treatments that can be considered, it is essential the dentist knows the patient's risk acceptance. If a management strategy involves some degree of unpleasantness, either during treatment or as a result of deciding not to treat a condition, the patient's degree of risk acceptance is an essential component when determining the best clinical decision.

Recognising patients' risk acceptances may help explain why patients sometimes appear to opt for courses of action which dentists would not recommend. For example, a dentist would probably suggest that a symptomless non-vital tooth should be root treated to avoid the possibility of abscesses and episodes of pain in the future. In contrast, a patient might (especially if no pain has been experienced) be prepared to accept the seemingly small risk of having to endure severe pain in the future in order to avoid the inconvenience of treatment at a time when no symptoms are present.

Patient involvement

A patient's risk acceptance and assessment of the value of a dental health state which results from treatment is an important part of the decision making process. Dentists should endeavour to determine these before embarking on any treatment strategy. Some dentists, however, may be too ready to adopt a personal viewpoint which can be out of step with that of their patients. To an extent this is unavoidable, but taking care to communicate the risks as accurately and clearly as possible, taking into account the patient's risk acceptance and value attached to the outcome, is clearly something to which dentists ought to aspire, if they want their patients to make informed choices.

There is a danger, when acting as a health care professional, of employing a somewhat paternalistic attitude. In the past, patients may merely have been informed of what treatment they should have, but it is now increasingly accepted that relationships between dentists and patients and patient satisfaction are improved by encouraging them to participate in choices about their own oral health and dental care. To achieve this type of mutually acceptable dialogue, dentists must know the probability of success for each type of treatment and be able to pass this information to the patient to enable them to make informed choices.

Conclusion

Thinking of disease as a simple disruption of anatomy and physiology, almost in the same way as we think of a machine breaking down, is no longer acceptable. Health is a feeling of well-being and control, as well as simply the absence of pathology. The roles of 'the sick' and 'the healer' for patient and dentist are no longer appropriate. If dentists are to fulfil their patients' needs and provide quality care, the process of making treatment decisions must take into account patients' social, psychological and functional needs as well as their disease status. In order to make the best treatment decision, dentists must use their

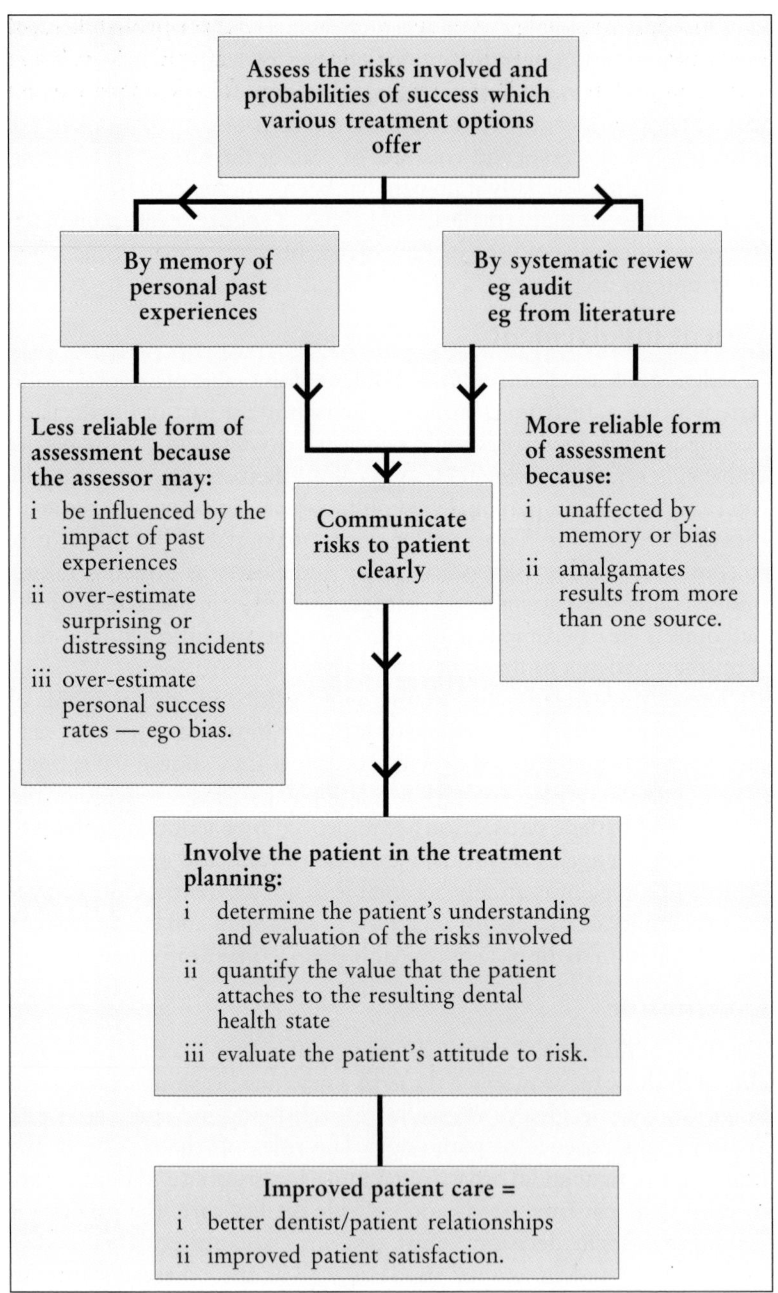

Fig. 4.2. The process of risk assessment in clinical decision making

expertise and knowledge to assess and communicate the probabilities of outcomes, while taking account of their patients' attitudes to risk and treatment preferences. The many issues involved in risk assessment are listed in figure 4.2.

Summary of chapter 4

Risks and probabilities

- **Assessing the probability of success**
  The probability of success or failure from a particular treatment can be assessed either intuitively from memory or by systematically reviewing the available data.

- **Using memory to assess probabilities**
  Memory can be deceptive. Firstly, particularly awful or challenging results are remembered more clearly than 'run of the mill' events. Events which are easily remembered tend to be thought of as events which occur frequently. Therefore, distressing and surprising results from treatments may be considered more probable than they actually are.
  Also, 'ego bias' means that we recall events which cause us to see ourselves in a positive light more easily than we recall events which make us feel bad about ourselves. Once again, these more easily remembered events tend to be thought of as occurring frequently and therefore the probability of their occurrence is over estimated.

- **Using systematic review to assess risk and probabilities**
  Systematic review is a more accurate way of assessing risks and probabilities. Particularly useful is a review of one's own performance in practice. This allows the dentist to give his patient highly accurate estimates of what will happen subsequent to treatment. However, frequently it is necessary to rely on the published literature. Although this is a better method of estimating success/failure rates from treatments than memory and guesswork, it is important to remember the potentially biasing effects of careful patient selection and specialist operators who are usually involved in research. The published results may not therefore be directly relevant to the situation in a general dental practice.

- **Explaining risks and probabilities**
  Whether words/phrases or numbers are the most suitable vehicle with which to present risk assessments to patients depends very much on the patient. The most important point is that neither words, phrases nor numbers are completely unambiguous and

the dentist must attempt to obtain feedback from the patient in order to assess the patient's comprehension of what has been said.

- **Using words and phrases to explain probabilities**
  Words and phrases are particularly prone to misinterpretation, especially if vague phrases such as 'poor chance' are used; perhaps in order to muddy the water when we are unsure about the true probability of events.

- **Using numbers to convey probabilities**
  Numbers, too, can sometimes cause difficulty as probabilities can be expressed as percentages, fractions and proportions — all of which may be understood differently by a patient. However, in general, you will convey far more meaning using a numerical expression than you will using a verbal phrase.

Case history
A patient presents to you with a failed root treatment which has been re-root treated twice over the past 6 years by another dentist. She has heard that there may be an operation which would save the tooth. She asks you whether she should have the 'operation' and whether you will do it. Your experience with apicectomies is very limited and the last one you did went horribly wrong. Consider what you might say to this patient.

Our suggestions
This patient is basically asking you for an assessment of the risks and benefits involved if she were to have an apicectomy. In order to allow her to make an informed choice about having the treatment, you need to supply her with the following information, in a way which she understands.

(i) what is the probability of success?
what is the probability of failure?

(ii) if the operation is successful, how long is the tooth likely to last?
if the operation is a failure, will the tooth have to be extracted, or would a second operation be possible?

If you rely on your memory, you might say that she has a 50:50 chance of success since you have recently had a very bad experience with a similar treatment and limited experience of success with the treatment in the past. However, the chances of success may be greater in more experienced hands. It might be worth contacting an oral surgery colleague who would be able to tell you the published rates of success and the 5 and 10 year survival rates of apicected teeth.

What is vital is that the patient makes her choice based on the best possible evidence. Misinforming patients, either out of ignorance of the facts or personal bias, is denying patients the right to make informed choices.

Points you may wish to consider

- How might the risks associated with, and probability of success of, a given treatment be assessed?
- What do you believe to be the most reliable way of calculating the risks and probabilities associated with a treatment?
- Would you consider words/phrases or numbers to be the best medium to convey a probability to a patient? Consider the advantages and disadvantages of each.

5

Patient preferences and their influence on decision making

Chapter 4 introduced the idea that the value a patient places on an outcome has an important influence on the decision making process. This chapter considers this concept further. It argues that dentists should assess a patient's preferences and consider these before deciding on a treatment option.

Introduction

In the last chapter we examined how the probability of occurrence of various dental health outcomes might be assessed. The chapter also explored the best ways to communicate these probabilities to the patient concerned. However, there is a second and vital piece of information which a dentist needs to make a rational treatment decision: the value that the patient places on each potential outcome.

Events which might influence the value placed on a particular outcome can be identified by enumerating all the possible outcomes of all the treatment options available. The dentist can then determine the most and least favourable treatment outcomes and thus the best treatment option. It is important to note that the best treatment option is one that not only involves the least risk and maximum probability of success but, perhaps even more importantly, one that leads to a result which the patient finds favourable.

This chapter describes the process of enumerating the possible outcomes and explains how the value placed on a resulting dental health state can affect the decisions made in the dental surgery. It also describes how to map the decision making process, from a treatment option to outcome, so that each treatment decision can be analysed closely.

Enumerating outcomes

It is sometimes tempting to consider only the outcomes which are

- all decisions (or courses of action) that could be taken
- everything else that could happen which could affect the final outcome (chance events)
- all possible outcomes which could arise as a result of making any particular decisions
- the likelihood that each chance event or outcome could happen
- a measure of the value of each outcome to the patient.

Fig. 5.1 Five sets of information required before making a decision

'wanted' rather than taking account of all the possible outcomes, even those which neither dentist nor patient would find favourable. It is important when enumerating treatment options that all possible outcomes, bad and good, are considered. A good dentist will take account of the possibility of poor outcomes as well as considering the preferred end result.

However, when enumerating all possible outcomes, there are some difficult issues to be considered. For example, is there any difference in the outcome from restoring a tooth with a static lesion, and restoring a tooth with an active lesion? Should the two outcomes be considered to be identical because both result in a filled tooth, or should they be classed as different because they achieve different things?

It is likely that most dentists would put different values on the two treatments because one could be considered to be unnecessary while the other eliminates active disease. The patient's psychological reaction to a particular treatment would also be affected by the knowledge that in one case it effectively controls the disease and in the other it merely obturates a cavity.

Therefore, the same end result (a filled tooth, for example) does not always mean that the same outcome is achieved. When enumerating outcomes it is vital to determine what the treatment aims to achieve, because this will affect the patient's and dentist's perception of the treatment and thus the value placed on the outcome.

Decision 'trees'

Having enumerated all the possible options and outcomes for a particular symptom or disease, five sets of information must be determined and considered before making a final treatment decision (see fig. 5.1). This process can be laid out as a tree or flow diagram in which decisions lead to outcomes. The structure of the tree is effectively a map of all the possible events which could happen, having taken a particular decision (see fig. 5.2).

Mapping out all the events that can occur is a useful exercise in formalising decision making itself. However, the principal reason for constructing a tree is to analyse a decision so that the optimum treatment strategy can be identified.

Valuing clinical outcomes

It is impossible to make 'good' clinical decisions without knowing both the likelihood and the value of the various outcomes which treatment may bring about. It is also important to remember that what the dentist considers to be a 'good' outcome may not be the same as what the patient considers to be favourable.

Dentists' and patients' treatment values can vary

A dentist may consider that a tooth saved by a three- or four-visit root treatment is advantageous to a patient. However, the patient may not consider the saving of a tooth to be worth the discomfort, cost and time off work which would be involved. Thus, when coming to a decision, all of the psychological, sociological, monetary and functional issues which relate to treatment, as well as the health issues, should be taken into account.

The value that a patient places on an outcome should always be taken into account. For example, fig. 5.3 illustrates the restore/leave/extract decision in the form of a decision tree. Although there are more potential events which might come into play and the tree is not exhaustive, its purpose is to highlight the different outcomes and the factors which might influence how the patient values the results. This

A hypothetical (but realistic) example of decision making: should I take an umbrella with me?

As an example of how decision trees are formed, assume you need to decide whether to take an umbrella on your journey to the surgery. Fig. 5.2 illustrates a decision tree you might construct to determine what the optimum decision would be. There are only two decisions available to you — either to take the umbrella or not.

There are two chance events over which you have no control but which will affect the final outcome: whether it rains or not and whether you lose your umbrella or not. In the decision model illustrated it has been assumed that it will generally tend to be sunny (on 90% of occasions) if you take an umbrella and generally rainy (on 99% of occasions) if you do not. Whilst this is not in accord with modern meteorological thinking it is perhaps no less accurate.

There is also a particular problem with making the decision to take an umbrella, which is that umbrellas are notoriously prone to being left on buses, park benches or any other place they are temporarily laid. In the decision model it is hypothesised that you have a 50% chance of losing the umbrella if it rains, but a much higher chance (90%) of losing it if it is sunny. This seems reasonable as it is far less likely that you will lose an umbrella if it is actually being used.

The tree terminates in a set of outcomes which could arise from the various combinations of the preceding events. In the case of this particular tree only two outcomes are considered to occur — either you keep dry or become wet. Keeping dry is considered to be the best possible state and as such it is assigned a value of 1.0. Becoming wet is a less favourable outcome and in the case of this example has a value of 0.5. If you follow the path marked out in blue in fig. 5.2, for example, you have decided to take an umbrella and it rains. However, because at some time in the interim you lost the umbrella, you got wet.

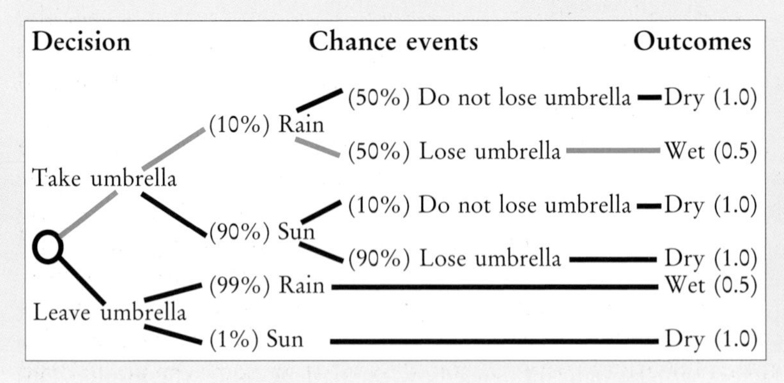

Fig. 5.2 A decision tree: should I take an umbrella with me?

enables the dentist to choose the best treatment option for that particular patient.

There are three possible treatment options available in the decision tree, but most dentists would agree that if a patient openly states that he or she prefers extraction to restoration, then removal of the tooth is the treatment of choice. In such a case, diagnostic skills are not of paramount importance, the patient's preferences are.

Assessing the value of clinical outcomes is the basis of good decision making and ensures that the patient receives what he or she considers to be 'value for money'. No treatment decision should be made without knowledge of the patient's preferences, because it is the patient who determines whether a treatment option is good or bad. Taking patient values into account before making a clinical decision will, therefore, improve the chances of the decision being a good one, as well as improving patient satisfaction.

Conclusions

The hypothetical examples in figs 5. 2 and 5. 3 are an attempt to explain the value judgements which are part of making 'good' clinical decisions. A 'good' decision raises the probability of reaching outcomes which, to the patient, have high value, whilst concomitantly lowering the probability of events occurring which (in the patient's eyes) have low value. It is therefore the ability to elicit from a patient the attitudes upon which an outcome evaluation depends, that is the mark of a truly skilled dentist.

Professional training, which enables us to accurately assess the probability of certain events occurring when pathologies are treated or not treated, is, of course, vital to treatment decision making. However, it is the ability to relate with patients in a way that enables them to communicate the values they place on outcomes, that makes dentists 'health care professionals' rather than simply highly skilled diagnosticians and technicians.

Although the aim of the dentist is to provide the best clinical care under a given set of circumstances, the patient's views regarding the preferred outcome is axiomatic to good dentistry. Patient preferences are an important part of clinical decision making and should never be overlooked when deciding upon the best treatment option.

It is therefore highly inappropriate to justify or rationalise treatment on the basis of the amount of pathology present. The optimal treatment plan is dictated by what outcome can be achieved and how valuable this is to the patient, not by the amount of disease

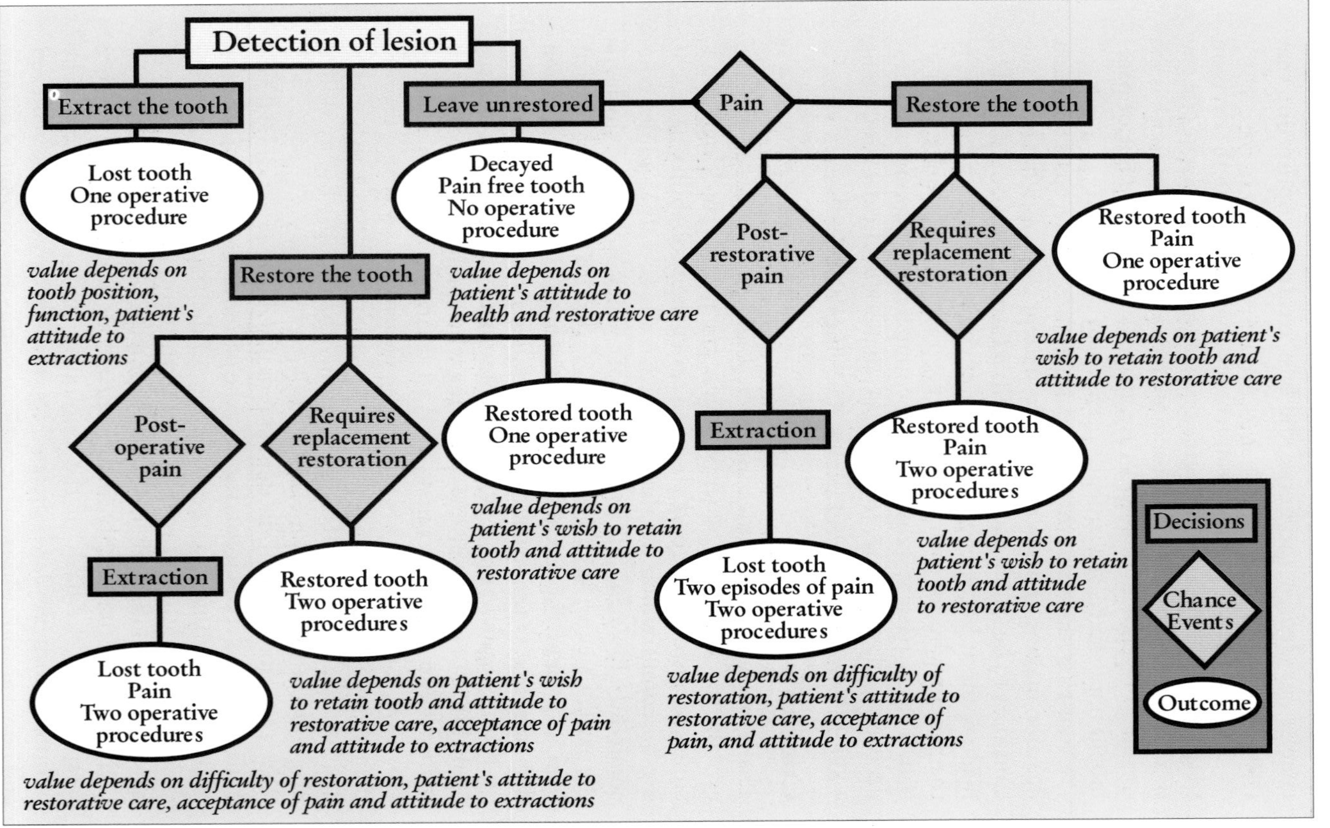

Fig. 5.3 A decision tree: restoration of carious lesions

A clinical example of decision making: should I restore this tooth

When faced with a tooth in which you believe there to be caries, you have three options (see fig. 5.3). First, you could extract the tooth — the outcome from this action is, in its simplest terms, a lost tooth. However, the loss of a dysfunctional tooth, or one which is painful, might be considered to be a 'better' outcome than the loss of one which is functional or non-painful at the time of the decision. Therefore, the value which the patient puts on the tooth, for whatever reasons, will affect the value of the outcome of your action.

Alternatively you might opt to leave the tooth unrestored. If the lesion is shallow, or if the patient's behaviour is such that the lesion will progress very slowly, the outcome from making a decision not to intervene is a decayed but pain-free tooth. However, another outcome from leaving the tooth alone might be that the tooth becomes painful after a year and then needs to be extracted or filled.

Similarly, if you leave the tooth unrestored it might become painful within the next week. If you fill it at that point, the outcome is the same as opting to fill now (a filled tooth), but the patient will have suffered pain (and may also call your diagnostic abilities into question). However, for someone who finds restorative treatment a very difficult and frightening experience, or is short of time or money, delaying action today may be a better option if you think that the probability of future problems is low.

Finally, you might opt to restore the tooth today. Although you might like to think that this action effects a 'cure' it is possible that the lesion is very deep and will become painful and need to be extracted following restorative care. In this case the outcome is a lost tooth plus restoration and prolonged pain. You also need to remember that a restoration commits the patient to a lifetime of repair and possible extension of the cavity, so the end outcome might be a filled, refilled tooth.

present at the outset (see fig. 5.4). Dentistry is therefore increasingly about holistic care of individuals, rather than the treatment of pathology in mouths.

- consider all possible decisions that can be made
- consider all chance events that may occur
- from these enumerate all possible outcomes
- determine the likelihood of all possible events, even chance events
- place a value on each outcome based on
    - i the likelihood of it occurring
    - ii the dentist's preferences
    - iii the patient's preferences
- when considering a patient's preferences, the following must be considered:
    - i value of aesthetic outcome
    - ii value of health outcome
    - iii time available
    - iv ability/willingness to spend money
- if all these are taken into account, a 'good' clinical decision can be reached.

Fig. 5.4 Process of making a 'good' clinical decision

Summary of chapter 5

Planning treatment and making decisions

- **What we need to know in order to make a treatment plan**

  Before embarking upon a particular course of treatment for a patient, the information needed is:

  (i) the patient's chief complaint or worry
  (ii) all the possible courses of action which might alleviate/obviate the patient's problem
  (iii) the events and circumstances (such as patient's diet, etc.) which might affect the final outcome
  (iv) the probability of success/failure of each potential course of action
  (v) the importance of the potential outcomes to the patient concerned

- **Decision trees**
  A decision tree is a graphical method of mapping out all the decisions, chance events and dental outcomes which might result from a decision. It allows the decision maker to calculate the best course of action given the patient's evaluation of each possible outcome. It gives guidance regarding which decision will have the highest probability of achieving the desired outcome.

Case History

The following patient presents at your surgery.

Mrs Beeton is a 37-year-old professional lady, with little time to spare. She has a well-paid job which involves PR work and public speaking. She presents to you with acute, hitherto intermittent, pain in ⌊6⌋. The tooth is extremely sensitive to heat and cold. Mrs Beeton cannot work because of the pain. Consider what treatment options you could offer and which you think would offer the patient the best solution to her problem.

Case history — our suggestions

Mrs Beeton's pain is clearly her main problem. The first step is to enumerate the possible courses of action for dealing with the patient's problem. These would be:

(1) extract the tooth
(2) place a sedative dressing in the tooth
(3) commence a root treatment

The second step would be to consider what events might affect the outcomes from each of these possible actions. These might include events such as tooth fracture during extraction, dry socket, no effect from dressing, fracture of tooth while dressed, inability to adequately debride the root canals, etc.

The third step is to determine the potential outcomes. If the tooth is extracted, these are:

(a) simple extraction, no difficulties or complications
(b) surgical extraction
(c) either of the above, plus post-operative pain/swelling

If the tooth is dressed, these are:

(a) reduced/no pain
(b) continued pain
(c) tooth becomes tender to percussion (worse)
(d) tooth fracture

If the tooth is root-treated, the treatment:

(a) may fail and the tooth will require extraction
(b) may succeed
(c) may succeed but not without some pain, etc. during treatment

The fourth step is to try to estimate how likely each of the chance events are — this will entail taking radiographs to assess root formation (for extraction or RCT), assessment of the efficacy of medicaments which can be applied to the tooth, assessment of success rates for all three types of treatment in the past, etc.

The final and most important step is to find out, from the patient, how she feels about each of the outcomes. For example, would the loss of a posterior tooth be of concern to Mrs Beeton? Does she value her tooth sufficiently to wish to pay for RCT? Indeed, does she have the time to spare to undergo molar endodontics, etc.?

From this information, although Mrs Beeton might place the highest value on the root-treated tooth, if your professional assessment is that the probability of success is low, then this may not represent the best option. Therefore, by weighting the value of each outcome by the probability of its occurrence, the best solution to the problem will be found.

6

Decision making in dental practice: a case study

This chapter illustrates in practical terms the theory described in previous chapters. An outline of the basic steps in decision making is used in order to illustrate the points raised.

Introduction

The previous chapters have looked at various factors which might influence the way in which dentists make decisions. This final chapter attempts, by way of a detailed case study, to pull together the strands involved in decision making, in order to give an overview of the information and thought processes which can, and should, influence treatment planning.

- thorough knowledge of all the possible treatment options
- comprehensive view of all possible consequences of each available treatment
- the ability to attach values to each possible outcome
- the ability to rank the values, from the most desired outcome to the least desired outcome

Fig. 6.1 What is needed to make a rational treatment decision.

In order to make good decisions on a rational basis, several specifications must be fulfilled (see fig. 6.1).

First, dentists need to have complete knowledge of all the possible alternative courses of action, given a particular pathology or clinical problem. It is assumed that this base of information is provided for the dentist during undergraduate training and updated through subsequent postgraduate education.

Second, the dentist needs to have a comprehensive view of all the possible consequences of each available treatment, including the undesirable outcomes. In order to make a good decision, a dentist must know the outcomes of all the treatment options, including new types of treatment, and must be able to make correct assessments of the likelihood of success or failure of each alternative treatment.

The third skill which dentists require to make treatment planning decisions is the attachment of 'values' or 'pay-offs' to each of the possible outcomes. That is, the dentist must be able to say how he or she and, most importantly, the patient values the different outcomes. Ideally this should be determined through discussion with the patient and perhaps by using the more formal evaluative techniques such as those discussed chapter 5.

Lastly, dentists must be able to order the value of each course of action in a sequence from the 'highest' to the 'lowest'. That is, the dentist must be able to say which outcome is the best and which is the worst. It is only then that a course of action can be decided upon which is most likely to lead to the outcome that the dentist has decided has the highest value both in treatment and patient terms.

Unfortunately, all of the information listed above is rarely available to the dentist when making treatment planning decisions because of the following factors. First, there is the problem of uncertainty. The dentist may not know about, or be able to undertake, all possible alternative courses of action. For example, some dentists prefer to refer all patients who require surgical treatment. Therefore they cannot assess the probability or value of outcomes which would result from opting for various alternative surgical treatments.

Second, the value of each course of action may be difficult to assess because the dentist may have many criteria by which the outcomes are valued. These might include the patient's views, the financial costs and consequences of the course of action, or the amount of time available (on either the patient's or the dentist's behalf).

Lastly, there are limitations on how the dentist processes all the information available. It is unlikely that a dentist is able to evaluate and compare all possible alternative actions. Researching and evaluating every possible alternative is impractical in most instances because of the limited time that can be spent on each decision for each individual patient.

Stages in the treatment planning process

Treatment decision making should proceed in several stages. These are:

1 Define the problem

The dentist must first talk to the patient and define the problem. This will entail finding out the patient's reason for attending, its significance and whether there are any other problems which the patient is unaware of.

The dentist then needs to distinguish between conditions which 'must' be treated and conditions which 'could' be treated. For example, the dentist may decide: 'I must remove this tooth in order to relieve the patient's pain, and I could enhance the patient's subsequent appearance by fitting a bridge'.

2 Gather information

Next the dentist must gather and analyse all relevant facts. This may include finding out about the patient's social history, the amount of time and money the patient is able or prepared to spend on retaining or treating the dentition, and the effort that the patient is willing to undergo in order to maintain oral health subsequent to treatment. It is important to remember that patients have a right to behave as they see fit, and to embark on treatments which require efforts beyond their capabilities will lead to poor outcomes.

3 Enumerate options

The dentist must consider all the possible alternative treatments. This is a natural extension to the previous stage of fact gathering and many possible solutions will arise. It remains a danger that the first feasible solution will be accepted by both dentist and patient, rather than a choice being made between a number of them.

4 Evaluate options

If a number of possible solutions rather than just one feasible solution have been identified, the dentist must then evaluate each of the alternative solutions to the problem, in terms of oral health objectives. This stage in the treatment planning process will involve discussion with the patient to find out what he or she truly wishes to achieve.

5 Select the best option

The final stage of treatment planning is to select the best alternative. The choice will be based on the information which is available and will often be a compromise once all the various factors have been considered. That is, the dentist will judge possible solutions against each other by examining:

- the risks compared with the expected gain
- whether the patient is prepared to make the amount of effort each alternative solution requires
- whether the patient can, or is willing to, afford the cost, time and perhaps in some cases discomfort, for each possible option
- whether the dentist and the patient have the necessary skills to provide and maintain oral health subsequent to carrying out each of the alternative treatments.

It is also important when planning treatment in this rational way to analyse the possible consequences of a bad decision. That is, the dentist must anticipate problems which may arise from choosing a particular option, and it is a wise dentist who has already considered how such problems may be successfully dealt with.

Once the best treatment option has been chosen, the decision has to be implemented. This will include negotiation with the patient in order to ensure that the task is taken to completion. The dentist must be able to communicate with the patient in such a way that the dentist can check that the decision or treatment plan is fully acceptable to the patient.

The advantages of rational decision making

To make a treatment plan, values or costs need to be assigned to the different outcomes. Using such a technique has several advantages. It focuses the dentist's thinking on the factors which truly influence the decision and helps to structure organised thought. Taking a rational approach to treatment planning (see fig. 6.2) helps to uncover hidden assumptions and ensures that all possible options are considered. Furthermore, planning treatment in this rational way helps to provide an effective vehicle for communicating the options to the patient. It also identifies where additional information is needed by either dentist or patient before a satisfactory treatment plan can be reached.

Although the framework given above provides a rationale for treatment planning and dental decision making, it is clear that the 'complete dentist' develops expertise through informal processes as well, for example from his or her experience of the patient's behaviour, or from past observations of the element of chance in the clinical environment.

Rational dental decision making

The formal processes suggested in earlier chapters as a means for rationalising dental decision making may be too laborious for individual treatment decisions. However, they point the way towards decision

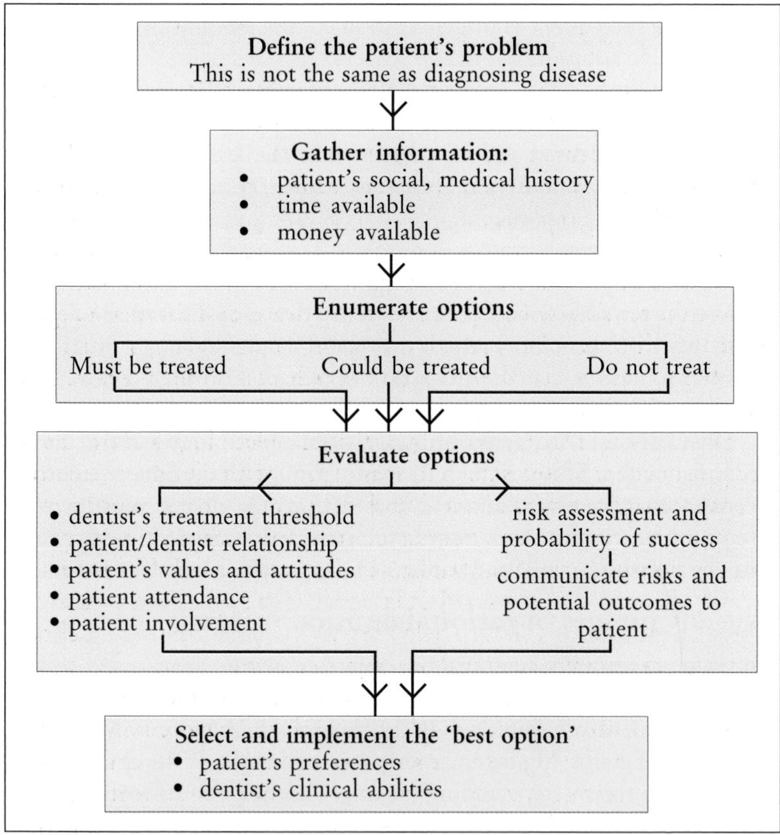

Fig. 6.2 Rational clinical decision making

making based on informed judgements and in which rational choices are made. The task of dentistry is focused on taking decisions about providing oral care for patients which will meet their needs and demands, and will match their lifestyles and behaviours. It is also about how to communicate and motivate patients.

Most decisions which dentists make are fairly routine and are encountered many times. However, some decisions are of a different nature and call for choosing between different, often risky, strategies in order to solve difficult or undefined patient problems. Rational decision making will help the dentist define the treatment options available, and the patient and dentist values involved, so that a decision can be made which will be successful not only from a clinical point of view but will also satisfy the patient.

Summary and conclusions

In the course of this book, we have concentrated on the theory underlying the decisions which are made in the dental surgery. Interest in this field has been rather neglected in the past, particularly by researchers. To examine the decision making process, we began by considering the complexity of the process and the many factors which are brought to bear, even subconsciously, on each clinical decision. Subsequently, we directed our attention to the sources of variation between practitioners and practices in the hope that, by defining the issues which make clinicians view a problem differently, it might be possible to rationalise the processes which come into play when a treatment decision is made.

This analysis led to the conclusion that oral pathology and treatment need are not simple correlates of one another but are simply two factors which inter-relate with a complex web of external and personal factors. They lead to dentists and patients coming together in an attempt to address the problems of the patient. Having recognised that treatment thresholds are not rigid points but indications of a spectrum of needs in which a dentist may be able to help the patient, quality care was defined — using patient satisfaction as the objective of care.

We have argued that risk assessment and communication with patients are central, rather than peripheral, to the decision making process and have highlighted the need for accurate and relevant research information on which practitioners may base their assessment of probabilities. Finally, taking all the above into account, we were left with a premise that patients' preferences, values and attitudes are of paramount importance during treatment decision making. This is a thesis which every successful practitioner could have told us at the outset — and yet perhaps they rarely disclaim this knowledge for fear that their judgements may then be considered unscientific.

In essence, this book has come to a conclusion which many others have arrived at by different methods. It is the patient who is the expert.

Hypothetical case history

The following case history looks at an example of clinical decision making. By using the framework of the various steps in the clinical decision making process, it clarifies how the theory of rational decision making is relevant to treatment planning in the dental surgery.

Peter is a 17-year-old, currently attending sixth-form college. He has attended your practice irregularly for 12 years. At the age of 5 he required the removal of several deciduous teeth because of rampant caries. At the age of 10 he had two carious lower first permanent molars removed.

He last attended the practice 2° years ago, when his oral hygiene was extremely poor and he was reluctant to take any advice or receive any treatment. He has now attended the practice, with no reported symptoms, requesting a 'check-up'. His oral hygiene has improved dramatically since his last attendance.

Define the problem

Most patients seek treatment because they perceive a problem or a potential problem rather than simply because they wish to be orally 'healthy'. Therefore, in Peter's case, it is essential to determine why he has attended the surgery at this particular time.

On questioning, you discover that Peter has recently acquired a girlfriend and this seems in some way to have acted as a prompt to his visit. He is not in pain but when questioned, says that he has a bad taste in his mouth in the mornings.

Through discussion you conclude that Peter merely wishes to know whether it is 'normal' for a person's mouth to taste less than fresh on rising, and that he is using this symptom to justify his dental visit to himself.

Gather information

On further questioning you recognise that Peter has reached a stage in his development whereby he has become very 'appearance conscious'. He is fit and well and trains regularly to stay in shape. His diet is reasonably well balanced, although he consumes approximately 3-4 cans of 'sports drinks' per day.

He is brushing his teeth regularly but possibly inadequately. He is worried about having 'bad breath' (his girlfriend has encouraged him to attend the surgery). He will not have to pay for any treatment and appears to be willing to undertake any treatment which you deem to be necessary.

On examination there are no overt signs of advanced caries and his oral hygiene is on the borderline of being acceptable. Periodontal assessment reveals nothing untoward. He has white spot lesions ($\underline{2|1}$) and radiographs reveal early approximal carious lesions on three teeth. Only one of these lesions appears to have reached the ADJ. Peter claims that he will attend the surgery on a regular basis from now on.

Enumerate options

Peter presents a problem in that most of the conditions which he exhibits 'could' rather than 'must' be treated. There are a number of options and many more exist, but for the purpose of simplicity the alternatives mentioned here are limited to a few of the most plausible (see fig. 6.2).

Evaluate options

After having enumerated the various options available to Peter, it is then your task to place a value on all the treatment options. An essential part of this process is talking to Peter to establish what he wants to achieve in terms of oral health. The value of the options will also be influenced by your own judgement of Peter's ability to maintain good oral hygiene and attend your surgery regularly. Your assessment of the probability of success of the various options will also determine whether an option is a good or bad one (see fig. 6.2).

Select the best option

It is clear from the above that selection of the best option relies heavily on the dentist's predictions of the patient's likely future behaviours, and his likely reactions to events. The key then to selecting the best option rests in the dentist's ability to communicate with, and glean all the relevant information about the patient's attitudes, habits and behaviours.

Enumerate the options

a You can tell Peter that he does not have any need for treatment and that you will see him again in 6 months' time. ('Doing nothing' should always be considered an option).

b You could explain to Peter that he must brush his teeth more thoroughly and that he must reduce the amount of sugared drinks he consumes, as he has some early decay. Then ask to see him again in 6 months' time, when you will review the progress of the carious lesions.

c You could ask Peter to attend your hygienist for intensive oral hygiene instruction.

d You could ask to see Peter yourself for oral hygiene instruction and general advice. You explain the carious process and demonstrate the lesions on his radiographs to him. You ask him to use a fluoride rinse and complete a diet diary. You see him on a monthly basis for 4 months.

e You could restore the tooth with the most advanced decay. Whilst doing so you explain the importance of keeping his mouth clean and reducing his sugar intake.

f You could restore all of the approximal lesions, apply fluoride varnish to the anterior teeth and tell Peter you will send for him in 1 year's time.

Evaluate the options

a If you take course of action *a*, Peter will be happy, as he will feel 'healthy'. Whether you take this course of action will depend on your assessment of the likelihood of progression of carious lesions within 6 months and your view as to how likely it is that Peter will re-attend. (He has, in his view, 'survived healthily' for 2° years without a dental visit. If you tell him that he is healthy, he is likely to assume that he will remain so, without assistance from you).

You are also not allowing Peter to participate in his own health. Essentially, you might take this course of action if you felt that Peter's oral health behaviour is such that lesion progression is highly improbable and nothing untoward will occur before you see him again. To choose this option you must also believe that Peter will attend for his next visit.

b If Peter shows fear of restorative treatment, and you feel that having explained his need to care for his mouth, he will be likely to remain on the border line of need for treatment, this may be a beneficial option. Whether or not it is will depend on how certain you are that Peter will re-attend. If you think that this is unlikely, the option has a lower value.

c Choosing option *c* may help to harness Peter's new found wish to care for his mouth. The benefits of this option depend on your assessment of how Peter will respond to your hygienist.

d Choosing option *d* has the potential to lead to a very good outcome. However, this depends on the probability of Peter adhering to the suggested regime. It is also possible that you are asking him to take in, understand and act upon too much information. Whether this opinion is the optimal one depends, therefore, on your assessment of Peter's ability to comprehend advice and explanation, and upon his current level of motivation.

e By undertaking restorative treatment of the most advanced decay you avoid the possibility of future pain for Peter if he does not re-attend. Before choosing this particular option you must be able to assess the probability of re-attendance, the likely rate of progression of the lesion, and the probability that the experience of restorative treatment will discourage or even encourage future attendance.

f This option may be the optimal one if your assessment of the patient indicates that he is unlikely to improve his oral health habits, and is highly unlikely to re-attend before extractions become necessary.

The dentist, of course, has access to specialist knowledge and has, importantly, thorough training and vast experience. However, treatment must be planned with a central theme that patients, through their experiences of life, culture and through discussion, have sets of ideas about what should happen to them. Both patient and dentist reach conclusions about the problem, the possible actions which could be taken, the process of treatment, etc., based on their own reasoning. Their reasoning may be flawed and based on misinformation, inconsistency and uncertainty. The issue for the dentist is to recognise and deal with the fact that uncertainty exists for the profession as well as the patient. We cannot continue to insist that all treatment plans are based on incontrovertible facts. Good treatment planning requires that the dentist places the patient in a situation where he can choose to take advantage of the dentist's specialist knowledge and skills — if the patient wishes to do so.

7

Practical exercises

This chapter contains some self-assessment exercises to help you systemise your own clinical decision making. Each set of exercises reviews issues covered in the corresponding chapters at the beginning of the book.

1 An introduction to clinical decision making

Questions

Q 1.1 Describe the factors other than levels of disease/clinical features which can and should influence treatment planning. Discuss how you would take these factors into account when deciding upon an appropriate course of treatment for an individual.

Q 1.2 Explain how the following might help to determine the optimum treatment plan for a patient.

(a) patient's age
(b) patient's dietary habits
(c) patient's risk attitude
(d) patient's lifestyle
(e) patient's occupation

Q 1.3 The following patient presents for treatment. Summarise the questions you would need to ask and the information you would need before making a definitive treatment plan.

Mr O'Neill is a 65-year-old man who has just retired from a lifetime in farming. He plans to emigrate in the near future. He has not attended a dentist for approximately 15 years.

Intra-orally, he has 19 standing teeth, none with restorations. All his molar teeth are missing, as is the $|\underline{2}$. His oral hygiene is reasonable although he has 3–4 mm recession on all upper teeth.

2 Making sense of treatment decisions

Questions

Q 2.1 Explain what is meant by the terms: 'perceptual variation' and 'judgemental variation'. Give a dentally related example of each of these types of variation and explain how each could be controlled. What is the easiest type of variation to control and why?

Q 2.2 If two dentists make different treatment plans for a patient, can we establish which of them is 'wrong'? What might cause the two dentists to differ?

Q 2.3 Find a suitable bitewing radiograph. Which tooth surfaces would you restore?

Compare your treatment plan with a colleague. If possible, make a consensus treatment plan, then compare this with one made by another colleague.

Discuss why there are discrepancies between the various plans. For example, did each observer see the same number of lesions?

Did each observer agree about the depth of the radiographic lesion?

Did the observers agree about the depth of the radiographic lesion but disagree about how radiographic appearance translates to clinical lesion depth?

If all observers agreed about the presence and depth of lesions, did they make similar treatment plans?

3 To treat or not to treat?

Questions

Q 3.1 Why is it that some dental practices offer completely different types of treatments for their patients when compared with others?

Q 3.2 Discuss what you understand by the terms:
(**a**) Treatment threshold
(**b**) Quality care

Q 3.3 Peter is a 15-year-old schoolboy who is a regular dental attender and has excellent oral hygiene. At a routine visit, you notice suspicious shadows beneath the occlusal surfaces of the lower first permanent molars. Clinically, the fissures appear deep and stained, but otherwise normal. You cannot perceive any signs of cavitation. What would you do?

4 Assessing risks and probabilities

Questions

Q 4.1 How might the risks and probabilities of success of a treatment be assessed? What is the most reliable way of calculating risks and probabilities associated with treatment?

Q 4.2 Should probabilities be presented to patients as words or as numbers? What are the problems with each method?

Q 4.3 A patient presents to you with a failed root treatment which has been re-root treated twice over the past six years by another dentist. She has heard that there may be an operation which would save the tooth. She asks you whether she should have the 'operation' and whether you will do it.

Your experience with apicectomies is very limited and the last one you did went horribly wrong. What would you say to the patient?

5 Patient preferences and their influence on decision making

Questions

Q 5.1 What information would you ideally like to have available to you before embarking on planning a course of treatment for a patient?

Q 5.2 What is a decision tree, and what is its purpose?

Q 5.3 Mrs Beeton is a 37-year-old professional lady, with little time to spare. She has a well-paid job which involves PR work and public speaking. She presents to you with acute, hitherto intermittent, pain in $\underline{6|}$. The tooth is extremely sensitive to heat and cold. Mrs Beeton cannot work because of the pain. What treatment options could you offer and which would you think would offer the patient the best solution to her problem?

Answers

A 1.1 (a) *Relationship with patient*

A patient may be either acquiescent or may demand what he or she wishes. To some extent the dentist must recognise the consumer voice and provide treatment which addresses the patient's wishes. On the other hand, with a patient who casts the dentist in an 'expert' role, it becomes difficult to determine the patient's true personal attitudes and values.

Without some understanding of what is of value to the patient, it is almost impossible to make realistic and appropriate decisions. Some participation from the patient is therefore a prerequisite to good treatment planning. It is a mistake to assume that the patient's value system is the same as the dentist's.

A 1.1 (b) *Patient personality*

The patient's personality will impinge on both the patient's value system and their attitudes to health, disease and, most particularly, risk. The patient's character will also affect how the dentist and patient interact with each other.

A 1.1 (c) *Patient's attendance pattern*

(i) the attendance pattern will, to some extent, reflect the patient's dental health attitudes and values. For example, you might assume that, if the patient attends the dentist for a check-up, that patient must be motivated and anxious to maintain healthy teeth despite the trouble and cost involved.

(ii) the attendance pattern may also give some indication of the patient's dental health behaviours. Patients who attend for regular check-ups and whose value systems make their oral health of high priority are likely to undertake behaviours which are conducive to health, ie they are likely to brush, floss and limit their sugar intake. Therefore, when planning treatment the probability of new disease and of disease progression can be expected to be lower.

(iii) Finally, regularity of attendance has an important influence on the setting of intervention thresholds. For example, if a patient presents with disease which is on the borderline of needing treatment, if that patient is a regular attender the dentist knows that the patient is likely to attend again before problems arise. Therefore, leaving disease untreated will not cause any untoward occurrences. On the other hand, because seeking regular check-ups indicates positive dental attitudes, the patient's expectations of dentistry may be higher. This may also influence the dentist's treatment threshold.

A 1.1 (d) *Costs and benefits*

For each individual patient, dental treatment and dental disease have a 'cost' attached to them. This does not mean only financial 'costs' but also costs in the sense of a psychological, emotional or social burden (eg who will look after the children while mum is having treatment). The 'costs' of disease often only become apparent to the patient when it is too late — pain has arisen. Patients will only take action if they view the consequences of their action as being of greater value than the cost. So, for some people, the knowledge that their teeth are in perfect health is worth all the effort required in order to make them so. Thus, a main considera- tion when planning treatment would be whether the benefits for the patient (the gain in health and well-being) are worth the expenditure (in money, time, angst, anxiety, discomfort).

A 1.2 (a) The patient's age affects both the probability of disease (90% of new caries occurs prior to age 15) and the likelihood and speed of disease progression. Similarly, age has an important effect on patients' behaviour, with younger patients mostly adhering to their parents' routines, teenagers 'doing their own thing', young single adults increasingly becoming sexually aware, and adults adopting the routines which fit most closely with their local culture and daily lifestyles.

Each of these considerations could and should influence which treatment plan suits an individual patient. In particular the type of advice offered, and the format in which it is presented must be tailored according to the patient's age.

A 1.2 (b) The patient's dietary patterns will affect the probability of caries and speed of caries progression. It may also reflect the patient's attitudes and values towards health and healthy behaviour and reveal his or her attitude to risks and benefits.

A 1.2 (c) A patient's attitude to risk will profoundly affect the treatment plan. For example, some mothers would far prefer their child to have teeth taken out under general anaesthetic (even though general anaesthetics are associated with a very small risk of death) than face the very high probability of their child finding a local anaesthetic extraction distressing. Mothers seem therefore to be more willing to take very small risks of a terrible outcome rather than a high risk of a mildly unpleasant outcome.

A 1.2 (d) Examination of the patient's current lifestyle may reveal his or her attitudes to risk, and will therefore give an indication of what is likely to be the optimum type of treatment plan.

A 1.2 (e) The patient's occupation may be a key to practical treatment planning. First, a patient's dental needs may be dictated by his or her occupation, eg does the patient perform in public at all? Singing, public speaking, playing a musical instrument will all affect the patient's dental needs, both in terms of the functional requirements and the aesthetic requirements.

Second, a patient's occupation will dictate the amount of time and money available for dental treatment. Closely supervised factory workers are less likely to be able to take paid time off work for dental treatment than professional office workers.

A 1.3 Question the patient regarding the reason for his sudden attendance. If he is symptomless, the prompt to the visit may be social (his daughter is getting married and he wishes the |2 replaced) or financial. He may be concerned about the mechanisms of acquiring dental care once he has emigrated, or perhaps he wishes to take advantage of the last NHS dentistry available to him. Alternatively, perhaps it is simply that, having retired, the patient has more time available for items such as attending the dentist.

Therefore, you will need to:

(i) ascertain whether the patient perceives any problems with his teeth

(ii) ascertain whether the patient feels any necessity to improve the state of his dentition

(iii) ascertain the time, money and effort the patient is prepared to commit to achieving any improvement.

For example, if his daughter's wedding is the motivation to attend in order to achieve an improvement in appearance, or if he simply wishes an 'overhaul' before emigrating, it would be inappropriate to offer Mr O'Neill a chrome cobalt denture with precision attachment stress breakers for the free-end saddles. Even if such an appliance is considered to be the last word in dental prosthetics —

(a) it is inappropriate to Mr O'Neill's needs and therefore the cost is likely to outweigh the benefit

(b) it would be expensive and difficult to maintain

(c) a number of lengthy procedures would be needed before the denture would be ready to be worn.

If the reasons for Mr O'Neill's attendance are social/aesthetic, perhaps a better solution would be a simple Maryland temporary bridge to replace $|\underline{2}$ (in time for the wedding if necessary). This would require minimal preparation, would not require a high level of maintenance, could be inexpensive and would address the patient's needs.

The best treatment for a patient is not necessarily the most complex and costly one.

A 2.1 Perceptual variation is the difference in opinion resulting from individuals perceiving things differently. An example would be two dentists viewing a radiograph, one with and one without a viewing box. The dentist using the viewing box would see many more lesions than the dentist who did not use a specialised light source.

It may be controlled by re-evaluating a case at a later date, seeking a second opinion, or using an alternative diagnostic system to confirm the diagnosis.

Judgemental variation occurs when dentists agree about what is seen, but have different views about how to treat it. For example, one dentist might believe an incomplete root filling indicates a need for re-root treatment, whilst another might believe that signs of inflammation would be required before such treatment is warranted. Judgemental variation may be controlled by the adoption of standardised clinical guidelines.

Perceptual variation is less easy to control because it tends to be random in nature and therefore difficult to predict when it will occur.

A 2.2 Two dentists will make different treatment plans for the same patient because of perceptual and judgemental differences. These in turn are influenced by patient and environmental factors.

The patient (including the clinical factors) may be perceived differently by each dentist. Their interpretation of the observations may then influence their judgement, ie their decisions about what would be best for the patient. If the dentists perceive different levels of disease, they may also view the patient differently. They are therefore highly likely to choose different treatment strategies.

It is not possible to determine who is 'wrong' overall. It may be possible to check the dentists' perceptions by reviewing cases etc., but each judgemental decision must remain unvalidated. Formal decision analysis (decision trees/sensitivity analysis) may, however, identify the most appropriate action to take for a given dental condition.

A 2.3 There is no single 'right' answer to this question but your discussions should have been instructive.

If the observers with whom you discussed the radiograph counted different numbers of lesions, the discrepancy arose from perceptual variation. Similarly, if the observers disagreed about the radiographic depth of the lesion, these differences again arise from differing perceptions.

In contrast, differences in opinion about the relationship between radiographic lesion depth and actual clinical lesion depth stem from judgemental variation. Finally, opinions about how best to treat a lesion of a given depth may vary. This, again, arises from judgemental variation.

A 3.1 There are two main explanations for practice variation. One theory is that the types of patient (and hence the type/pattern of disease) visiting one practice are different from the types of patient visiting another. Therefore, the patients' needs (in physiological, psychosocial and pathological terms) differ between practices. Therefore, if the dentists working therein are to adequately meet their patients' needs, they will offer different types of treatment.

The second common explanation of practice variation suggests it is simply that dentists choose to behave differently. This explanation implies either that there is a 'best' or 'correct' treatment for each type of disease, but that not all dentists manage to make that optimal choice, or alternatively, that no one actually knows what is the right or best treatment and that the choices made by dentists are fairly arbitrary.

Although both theories have some truth, neither completely or satisfactorily explains why practice profiles differ as widely as they do. Although practices may attract dissimilar patient pools, it is well known that dentists' practice profiles and treatment strategies can differ even when faced with very similar patients. These variations stem from the fact that the relationship between treatment need and the amount of disease present is not a simple linear one but is influenced by many external factors.

A 3.2 (a) The treatment threshold is the point in the progression of a disease at which inaction/non-treatment is perceived by the clinician as being likely to lead to a poor outcome for the patient. However, because clinical cases have 'grey areas' and because of the many external factors which impinge on decisions, it is not possible to define an exact point when treatment is, or is not, necessary. Therefore, although the terms treatment threshold or treatment criterion are commonly used, in actual fact the 'thresholds' used by clinicians are not rigid and definite points. Some dentists will set treatment thresholds at a level which ensures they never intervene except when there is no other possible course of action. Others will treat patients who they think 'have a good chance' of benefiting. Finally, some clinicians opt to treat all cases except those where there is absolute and incontrovertible evidence that no disease or pathology exists. Furthermore, one dentist may operate at either extreme, depending on the type of patient he or she is dealing with.

A 3.2 **(b)** Quality care involves making decisions to take actions which will increase the probability of an outcome or process which the patient regards as favourable while, at the same time, reducing the probability of an outcome or process which the patient regards as unfavourable.

Whether the outcome is regarded as favourable or unfavourable will vary according to the attitudes, behaviour and preferences of the patient, ie quality is about meeting the patient's requirements.

A 3.3 This is a dilemma in which many clinicians have found themselves. Perhaps the simplest way to approach the problem is to list the advantages and disadvantages of inaction or treatment (see fig. 7.1). It is important to bear in mind that these lists would read differently if the patient under consideration did not seek regular checkups.

A 4.1 The probability of success or failure from a particular treatment can be assessed either intuitively from memory or by systematically reviewing the available data.

Memory can be deceptive. First because, particularly bad or challenging results are remembered more clearly than 'run of the mill' events. Events which are easily remembered tend to be thought of as events which occur frequently. Therefore, distressing and surprising results from treatments will be likely to be considered more probable than they actually are.

Second because 'ego bias' means that we tend to recall events which cause us to see ourselves in a positive light more easily than we recall events which make us feel bad about ourselves. Once again, these more easily remembered events tend to be thought of as occurring frequently and therefore the probability of their occurrence is over estimated.

Systematic review is a more accurate way of assessing risks and probabilities. What is particularly useful is a review of your own performance in practice. This allows you to give your patient highly accurate estimates of what will happen subsequent to treatment. However, more frequently it is necessary to rely on the published literature. Although this is a better method of estimating success/failure rates from treatments than memory and guesswork, it is important to remember the potentially biasing effects of the careful patient selection and specialist operators who are usually involved in research. The published results may not therefore be directly relevant to the situation in general dental practice.

Advantages of restoring tooth	*Advantages of leaving a tooth unrestored*
Caries progression stopped	Tooth remains completely healthy if no caries is present.
Pain unlikely	You have avoided unnecessarily removing tooth tissue
Patient experiences restorative care before symptoms arise	The patient is happy
	You will have an opportunity to observe progress (if any) at next visit
Certainty that first permanent molars are not diseased	Patient does not have to have restorative treatment
The caries may be deeper than it appears and thus potential pulpal involvement will have been averted	Even if caries is present, its progression is slow enough for you to observe progression in a patient for regular checkups before opting to restore patient who goes for regular checkups

This dilemma represents the grey area of treatment planning. It is impossible to know for definite the extent and rate of progression of any caries in Peter's teeth since radiographs are somewhat unreliable for predicting the extent of occlusal caries. The treatment decision therefore rests upon whether a false positive decision (an unnecessary restoration) represents a better or worse outcome than a false negative decision.

Fig. 7.1 Advantages of restoring or leaving a tooth unrestored

A 4.2 Whether words/phrases or numbers are the most suitable vehicle with which to present risk assessments depends on the patient. The most important point is that neither words, phrases nor numbers are completely unambiguous and you must attempt to obtain feedback from your patients in order to assess their comprehension of what has been said.

Words and phrases are particularly prone to misinterpretation, especially if vague phrases such as 'poor chance' are used; perhaps in order to muddy the water when we are unsure about the true probability of events.

Numbers, too, can sometimes cause difficulty as probabilities can be expressed as percentages, fractions and proportions — all of which may be understood differently by a patient. However, in general, you will convey far more meaning using a numerical expression than you will using a verbal phrase.

A 4.3 This patient is basically asking you for an assessment of the risks and benefits involved if she were to have an apicectomy. In order to allow the lady to make an informed choice about having the treatment, you need to supply her with the following information, in a way which she understands.

(i) what is the probability of success? What is the probability of failure?

(ii) if the operation is successful, how long is the tooth likely to last? If the operation is a failure, will the tooth have to be extracted, or would a second operation be possible?

If you rely on your memory, you might say that she has a 50:50 chance of success since you have recently had a very bad experience with a similar treatment and limited experience of success with the treatment in the past. However, the chances of success may be greater in more experienced hands. It might be worth contacting an oral surgery colleague who would be able to tell you the published rates of success and the 5- and 10-year survival rates of apicected teeth.

What is vital is that the patient makes her choice based on the best possible evidence. Misinforming patients, either out of ignorance of the facts or personal bias, is denying your patients the right to make informed choices.

A 5.1 Before embarking on a particular course of treatment for a patient, the information you would need is:

(1) the patient's chief complaint or worry
(2) all the possible courses of action which might alleviate/obviate the patient's problem
(3) the events and circumstances (such as patient's diet, etc.) which might affect the final outcome
(4) the probability of success/failure of each potential course of action
(5) the importance of the potential outcomes to the patient concerned.

A 5.2 A decision tree is a graphical method of mapping out all the decisions, chance events and dental outcomes which might result from a decision. It allows the decision maker to calculate the best course of action given the patient's evaluation of each possible outcome. It will give guidance regarding the decision which will have the highest probability of achieving the desired outcome.

A 5.3 Mrs Beeton's pain is clearly her main problem. The first step is to enumerate the possible courses of action for dealing with the patient's problem. These would be:

(1) extract the tooth
(2) place a sedative dressing in the tooth
(3) commence a root treatment.

The second step would be to consider what events might affect the outcomes from each of these possible actions. These might include events such as tooth fracture during extraction, dry socket, no effect from dressing, fracture of tooth while dressed, inability to adequately debride the root canals, etc.

The third step is to determine the potential outcomes. If the tooth is extracted, these are:

(a) simple extraction, no difficulties or complications
(b) surgical extraction.
(c) either of the above, plus post-operative pain/swelling.

If the tooth is dressed, these are:

(a) reduced/no pain
(b) continued pain
(c) tooth becomes tender to percussion (worse)
(d) tooth fracture.

If the tooth is root-treated, the treatment:

(a) may fail and the tooth will require extraction
(b) may succeed
(c) may succeed but not without some pain, etc. during treatment.

The fourth step is to try to estimate how likely each of the chance events are — this will entail taking radiographs to assess root formation (for extraction or RCT), assessment of the efficacy of medicaments which can be applied to the tooth, assessment of success rates for all three types of treatment in the past, etc.

The final and most important step is to find out, from the patient, how she feels about each of the outcomes. For example, would the loss of a posterior tooth be of concern to Mrs Beeton? Does she value her tooth sufficiently to wish to pay for RCT? Indeed, does she have the time to spare to undergo molar endodontics?

From this information, although Mrs Beeton might place the highest value on the root treated tooth, if your professional assessment is that the probability of success is low, then this may not represent the best option. However, by weighing the value of each outcome by the probability of its occurrence, the best solution to the problem will be found.

Index